A Text-Book of Astronomy

CENTURY BOUND

Printed or published to the highest ethical standard

A TEXT-BOOK OF ASTRONOMY

By George C. Comstock

Published by Century Bound

First Published
1901

CONTENTS

Preface

The present work is not a compendium of astronomy or an outline course of popular reading in that science. It has been prepared as a text-book, and the author has purposely omitted from it much matter interesting as well as important to a complete view of the science, and has endeavored to concentrate attention upon those parts of the subject that possess special educational value. From this point of view matter which permits of experimental treatment with simple apparatus is of peculiar value and is given a prominence in the text beyond its just due in a well-balanced exposition of the elements of astronomy, while topics, such as the results of spectrum analysis, which depend upon elaborate apparatus, are in the experimental part of the work accorded much less space than their intrinsic importance would justify.

Teacher and student are alike urged to magnify the observational side of the subject and to strive to obtain in their work the maximum degree of precision of which their apparatus is capable. The instruments required are few and easily obtained. With exception of a watch and a protractor, all of the apparatus needed may be built by any one of fair mechanical talent who will follow the illustrations and descriptions of the text. In order that proper opportunity for observations may be had, the study should be pursued during the milder portion of the year, between April and November in northern latitudes, using clear weather for a direct study of the sky and cloudy days for book work.

The illustrations contained in the present work are worthy of as careful study as is the text, and many of them are intended as an aid to experimental work and accurate measurement, e. g., the star maps, the diagrams of the planetary orbits, pictures of the moon, sun, etc. If the school possesses a projection lantern, a set of astronomical slides to be used in connection with it may be made of great advantage, if the pictures are studied as an auxiliary to Nature. Mere display and scenic effect are of little value.

A brief bibliography of popular literature upon astronomy may be found at the end of this book, and it will be well if at least a part of these works can be placed in the school library and systematically used for supplementary reading. An added interest may be given to the study if one or more of the popular periodicals which deal with astronomy are taken regularly by the school and kept within easy reach of the students. From time to time the teacher may well assign topics treated in these periodicals to be read by individual students and presented to the class in the form of an essay.

The author is under obligations to many of his professional friends who have contributed illustrative matter for his text, and his thanks are in an especial manner due to the editors of the Astrophysical Journal, Astronomy and Astrophysics, and Popular Astronomy for permission to reproduce here plates which have appeared in those periodicals, and to Dr. Charles Boynton, who has kindly read and criticised the proofs.

GEORGE C. COMSTOCK.
UNIVERSITY OF WISCONSIN,
February, 1901.

A TOTAL SOLAR ECLIPSE.
After Burckhalter's photographs of the eclipse of May 28, 1900.

CHAPTER I.
DIFFERENT KINDS OF MEASUREMENT

1. ***Accurate measurement.***— Accurate measurement is the foundation of exact science, and at the very beginning of his study in astronomy the student should learn something of the astronomer's kind of measurement. He should practice measuring the stars with all possible care, and should seek to attain the most accurate results of which his instruments and apparatus are capable. The ordinary affairs of life furnish abundant illustration of some of these measurements, such as finding the length of a board in inches or the weight of a load of coal in pounds and measurements of both length and weight are of importance in astronomy, but of far greater astronomical importance than these are the measurement of angles and the measurement of time. A kitchen clock or a cheap watch is usually thought of as a machine to tell the "time of day," but it may be used to time a horse or a bicycler upon a race course, and then it becomes an instrument to measure the amount of time required for covering the length of the course. Astronomers use a clock in both of these ways— to tell the time at which something happens or is done, and to measure the amount of time required for something; and in using a clock for either purpose the student should learn to take the time from it to the nearest second or better, if it has a seconds hand, or to a small fraction of a minute, by estimating the position of the minute hand between the minute marks on the dial. Estimate the fraction in tenths of a minute, not in halves or quarters.

EXERCISE 1.— If several watches are available, let one person tap sharply upon a desk with a pencil and let each of the others note the time by the minute hand to the nearest tenth of a minute and record the observations as follows:

2h. 44.5m.	First tap.	2h. 46.4m.	1.9m.
2h. 44.9m.	Second tap.	2h. 46.7m.	1.8m.
2h. 46.6m.	Third tap.	2h. 48.6m.	2.0m.

The letters h and m are used as abbreviations for hour and minute. The first and second columns of the table are the record made by one student, and second and third the record made by another. After all the observations have been made and recorded they should be brought together and compared by taking the differences between the times recorded for each tap, as is shown in the last column. This difference shows how much faster one watch is than the other, and the agreement or disagreement of these differences shows the degree of accuracy of the observations. Keep up this practice until tenths of a minute can be estimated with fair precision.

2. ***Angles and their use.***— An angle is the amount of opening or difference of direction between two lines that cross each other. At twelve o'clock the hour and minute hand of a watch point in the same direction and the angle between them is zero. At one o'clock the minute hand is again at XII, but the hour hand has moved to I, one twelfth part of the circumference of the dial, and the angle between the hands is one twelfth of a circumference. It is customary to imagine the circumference of a dial to be cut up into 360 equal parts— i. e., each minute space of an ordinary dial to be subdivided into six equal parts, each of which is called a degree, and the measurement of an angle consists in finding how many of these degrees are included in the opening between its sides. At one o'clock the angle between the hands of a watch is thirty degrees, which is usually written 30°, at three o'clock it is 90°, at six o'clock 180°, etc.

A watch may be used to measure angles. How? But a more convenient instrument is the protractor, which is shown in Fig. 1, applied to the angle *A B C* and showing that *A B C* = 85° as nearly as the protractor scale can be read.

The student should have and use a protractor, such as is furnished with this book, for the numerous exercises which are to follow.

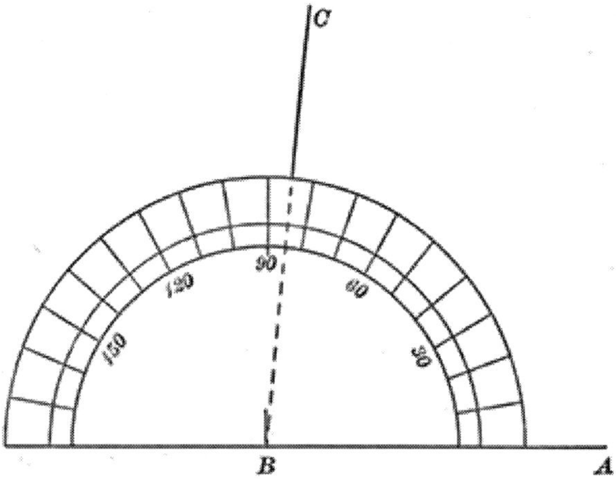

FIG. 1.—A protractor.

EXERCISE 2.— Draw neatly a triangle with sides about 100 millimeters long, measure each of its angles and take their sum. No matter what may be the shape of the triangle, this sum should be very nearly 180°— exactly 180° if the work were perfect— but perfection can seldom be attained and one of the first lessons to be learned in any science which deals with measurement is, that however careful we may be in our work some minute error will cling to it and our results can be only approximately correct. This, however, should not be taken as an excuse for careless work, but rather as a stimulus to extra effort in order that the unavoidable errors may be made as small as possible. In the present case the measured angles may be improved a little by adding (algebraically) to each of them one third of the amount by which their sum falls short of 180°, as in the following example:

	Measured angles.	*Correction*	*Corrected angles.*
	°	°	°
A	73.4	+ 0.1	73.5
B	49.3	+ 0.1	49.4
C	57.0	+ 0.1	57.1
Sum	179.7		180.0
Defect	+ 0.3		

This process is in very common use among astronomers, and is called "adjusting" the observations.

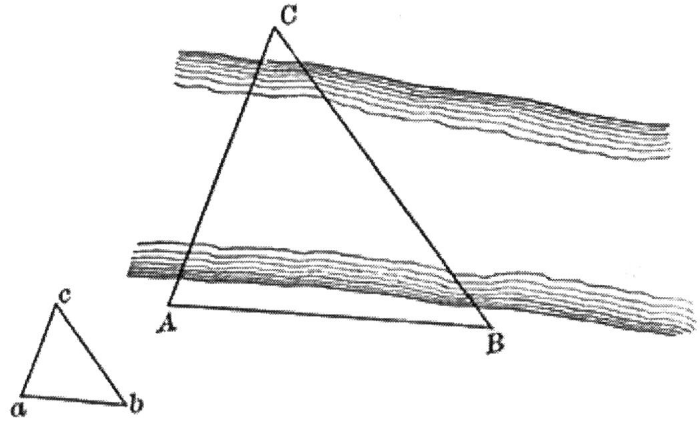

FIG. 2.— Triangulation.

3. ***Triangles.***— The instruments used by astronomers for the measurement of angles are usually provided with a telescope, which may be pointed at different objects, and with a scale, like that of the protractor, to measure the angle through which the telescope is turned in passing from one object to another. In this way it is possible to measure the angle between lines drawn from the instrument to two distant objects, such as two church steeples or the sun and moon, and this is usually called the angle between the objects. By measuring angles in this way it is possible to determine the distance to an inaccessible point, as shown in Fig. 2. A surveyor at *A* desires to know the

distance to *C*, on the opposite side of a river which he can not cross. He measures with a tape line along his own side of the stream the distance *A B* = 100 yards and then, with a suitable instrument, measures the angle at *A* between the points *C* and *B*, and the angle at *B* between *C* and *A*, finding *B A C* = 73.4°, *A B C* = 49.3°. To determine the distance *A C* he draws upon paper a line 100 millimeters long, and marks the ends *a* and *b*; with a protractor he constructs at *a* the angle *b a c* = 73.4°, and at *b* the angle *a b c* = 49.3°, and marks by *c* the point where the two lines thus drawn meet. With the millimeter scale he now measures the distance *a c* = 90.2 millimeters, which determines the distance *A C* across the river to be 90.2 yards, since the triangle on paper has been made similar to the one across the river, and millimeters on the one correspond to yards on the other. What is the proposition of geometry upon which this depends? The measured distance *A B* in the surveyor's problem is called a base line.

EXERCISE 3.– With a foot rule and a protractor measure a base line and the angles necessary to determine the length of the schoolroom. After the length has been thus found, measure it directly with the foot rule and compare the measured length with the one found from the angles. If any part of the work has been carelessly done, the student need not expect the results to agree.

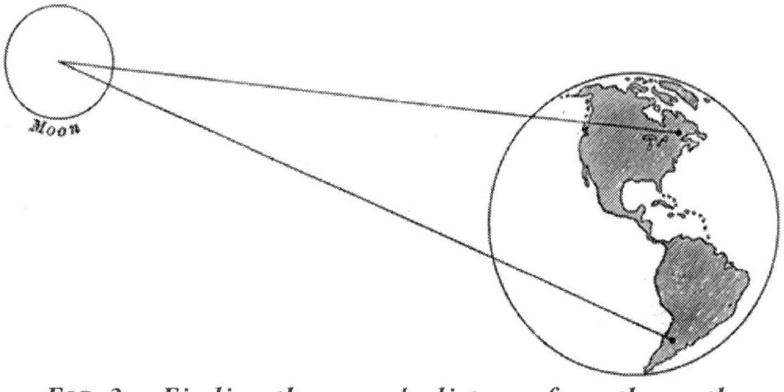

FIG. 3.— Finding the moon's distance from the earth.

In the same manner, by sighting at the moon from widely different parts of the earth, as in Fig. 3, the moon's distance from us is found to be about a quarter of a million miles. What is the base line in this case?

4. ***The horizon— altitudes.***— In their observations astronomers and sailors make much use of the *plane of the horizon*, and practically any flat and level surface, such as that of a smooth pond, may be regarded as a part of this plane and used as such. A very common observation relating to the plane of the horizon is called "taking the sun's altitude," and consists in measuring the angle between the sun's rays and the plane of the horizon upon which they fall. This angle between a line and a plane appears slightly different from the

angle between two lines, but is really the same thing, since it means the angle between the sun's rays and a line drawn in the plane of the horizon toward the point directly under the sun. Compare this with the definition given in the geographies, "The latitude of a point on the earth's surface is its angular distance north or south of the equator," and note that the latitude is the angle between the plane of the equator and a line drawn from the earth's center to the given point on its surface.

A convenient method of obtaining a part of the plane of the horizon for use in observation is as follows: Place a slate or a pane of glass upon a table in the sunshine. Slightly moisten its whole surface and then pour a little more water upon it near the center. If the water runs toward one side, thrust the edge of a thin wooden wedge under this side and block it up until the water shows no tendency to run one way rather than another; it is then level and a part of the plane of the horizon. Get several wedges ready before commencing the experiment. After they have been properly placed, drive a pin or tack behind each one so that it may not slip.

5. ***Taking the sun's altitude.*** EXERCISE 4.– Prepare a piece of board 20 centimeters, or more, square, planed smooth on one face and one edge. Drive a pin perpendicularly into the face of the board, near the middle of the planed edge. Set the board on edge on the horizon plane and turn it edgewise toward the sun so that a shadow of the pin is cast on the plane. Stick another pin into the board, near its upper edge, so that its shadow shall fall exactly upon the shadow of the first pin, and with a watch or clock observe the time at which the two shadows coincide. Without lifting the board from the plane, turn it around so that the opposite edge is directed toward the sun and set a third pin just as the second one was placed, and again take the time. Remove the pins and draw fine pencil lines, connecting the holes, as shown in Fig. 4, and with the protractor measure the angle thus marked. The student who has studied elementary geometry should be able to demonstrate that at the mean of the two recorded times the sun's altitude was equal to one half of the angle measured in the figure.

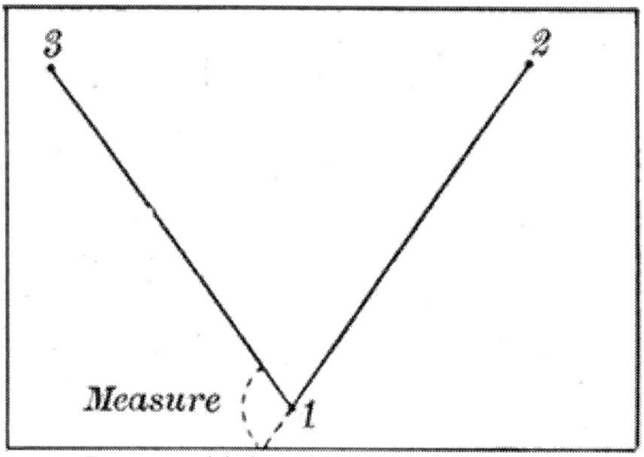

FIG. 4.— Taking the sun's altitude.

When the board is turned edgewise toward the sun so that its shadow is as thin as possible, rule a pencil line alongside it on the horizon plane. The angle which this line makes with a line pointing due south is called the sun's *azimuth*. When the sun is south, its azimuth is zero; when west, it is 90°; when east, 270°, etc.

EXERCISE 5.– Let a number of different students take the sun's altitude during both the morning and afternoon session and note the time of each observation, to the nearest minute. Verify the setting of the plane of the horizon from time to time, to make sure that no change has occurred in it.

6. ***Graphical representations.*** – Make a graph (drawing) of all the observations, similar to Fig. 5, and find by bisecting a set of chords *g* to *g*, *e* to *e*, *d* to *d*, drawn parallel to *B B*, the time at which the sun's altitude was greatest. In Fig. 5 we see from the intersection of *M M* with *B B* that this time was 11h. 50m.

The method of graphs which is here introduced is of great importance in physical science, and the student should carefully observe in Fig. 5 that the line *B B* is a scale of times, which may be made long or short, provided only the intervals between consecutive hours 9 to 10, 10 to 11, 11 to 12, etc., are equal. The distance of each little circle from *B B* is taken proportional to the sun's altitude, and may be upon any desired scale– e. g., a millimeter to a degree– provided the same scale is used for all observations. Each circle is placed accurately over that part of the base line which corresponds to the time at which the altitude was taken. Square ruled paper is very convenient, although not necessary, for such diagrams. It is especially to be noted that from the few observations which are represented in the figure a smooth curve has been drawn through the circles which represent the sun's altitude, and this curve shows the altitude of the sun at every moment between 9 A. M. and 3 P. M. In Fig. 5 the sun's altitude at noon was 57°. What was it at half past two?

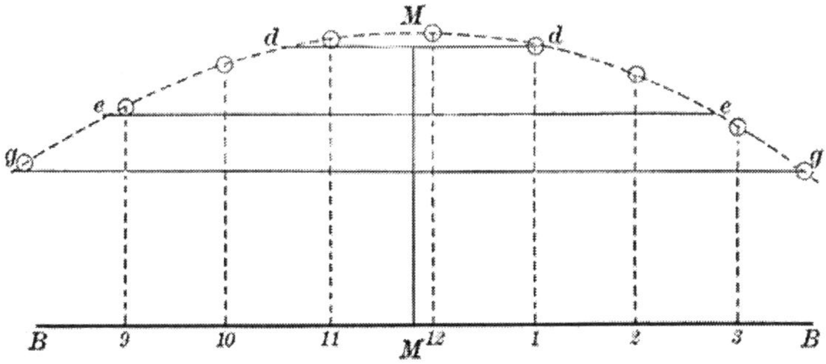

FIG. 5.—A graph of the sun's altitude.

7. ***Diameter of a distant object.***— By sighting over a protractor, measure the angle between imaginary lines drawn from it to the opposite sides of a window. Carry the protractor farther away from the window and repeat the experiment, to see how much the angle changes. The angle thus measured is called "the angle subtended" by the window at the place where the measurement was made. If this place was squarely in front of the window we may draw upon paper an angle equal to the measured one and lay off from the vertex along its sides a distance proportional to the distance of the window— e. g., a millimeter for each centimeter of real distance. If a cross line be now drawn connecting the points thus found, its length will be proportional to the width of the window, and the width may be read off to scale, a centimeter for every millimeter in the length of the cross line.

The astronomer who measures with an appropriate instrument the angle subtended by the moon may in an entirely similar manner find the moon's diameter and has, in fact, found it to be 2,163 miles. Can the same method be used to find the diameter of the sun? A planet? The earth?

CHAPTER II.
THE STARS AND THEIR DIURNAL MOTION

8. ***The stars.***— From the very beginning of his study in astronomy, and as frequently as possible, the student should practice watching the stars by night, to become acquainted with the constellations and their movements. As an introduction to this study he may face toward the north, and compare the stars which he sees in that part of the sky with the map of the northern heavens, given on Plate I, opposite page 124. Turn the map around, upside down if necessary, until the stars upon it match the brighter ones in the sky. Note how the stars are grouped in such conspicuous constellations as the Big Dipper (Ursa Major), the Little Dipper (Ursa Minor), and Cassiopeia. These three constellations should be learned so that they can be recognized at any time.

The names of the stars.— Facing the star map is a key which contains the names of the more important constellations and the names of the brighter stars in their constellations. These names are for the most part a Greek letter prefixed to the genitive case of the Latin name of the constellation. (See the Greek alphabet printed at the end of the book.)

9. ***Magnitudes of the stars.***— Nearly nineteen centuries ago St. Paul noted that "one star differeth from another star in glory," and no more apt words can be found to mark the difference of brightness which the stars present. Even prior to St. Paul's day the ancient Greek astronomers had divided the stars in respect of brightness into six groups, which the modern astronomers still use, calling each group a *magnitude*. Thus a few of the brightest stars are said to be of the first magnitude, the great mass of faint ones which are just visible to the unaided eye are said to be of the sixth magnitude, and intermediate degrees of brilliancy are represented by the intermediate magnitudes, second, third, fourth, and fifth. The student must not be misled by the word magnitude. It has no reference to the size of the stars, but only to their brightness, and on the star maps of this book the larger and smaller circles by which the stars are represented indicate only the brightness of the stars according to the system of magnitudes. Following the indications of these maps, the student should, in learning the principal stars and constellations, learn also to recognize how bright is a star of the second, fourth, or other magnitude.

10. ***Observing the stars.***— Find on the map and in the sky the stars α Ursæ Minoris, α Ursæ Majoris, β Ursæ Majoris. What geometrical figure will fit on to these stars? In addition to its regular name, α Ursæ Minoris is frequently called by the special name Polaris, or the pole star. Why are the other two stars called "the Pointers"? What letter of the alphabet do the five bright stars in Cassiopeia suggest?

EXERCISE 6.– Stand in such a position that Polaris is just hidden behind the corner of a building or some other vertical line, and mark upon the key map as accurately as possible the position of this line with respect to the other stars, showing which stars are to the right and which are to the left of it. Record the time (date, hour, and minute) at which this observation was made. An hour or two later repeat the observation at the same place, draw the line and note the time, and you will find that the line last drawn upon the map does not agree with the first one. The stars have changed their positions, and with respect to the vertical line the Pointers are now in a different direction from Polaris. Measure with a protractor the angle between the two lines drawn in the map, and use this angle and the recorded times of the observation to find how many degrees per hour this direction is changing. It should be about 15° per hour. If the observation were repeated 12 hours after the first recorded time, what would be the position of the vertical line among the stars? What would it be 24 hours later? A week later? Repeat the observation on the next clear night, and allowing for the number of whole revolutions made by the stars between the two dates, again determine from the time interval a more accurate value of the rate at which the stars move.

The motion of the stars which the student has here detected is called their "diurnal" motion. What is the significance of the word diurnal?

In the preceding paragraph there is introduced a method of great importance in astronomical practice– i. e., determining something– in this case the rate per hour, from observations separated by a long interval of time, in order to get a more accurate value than could be found from a short interval. Why is it more accurate? To determine the rate at which the planet Mars rotates about its axis, astronomers use observations separated by an interval of more than 200 years, during which the planet made more than 75,000 revolutions upon its axis. If we were to write out in algebraic form an equation for determining the length of one revolution of Mars about its axis, the large number, 75,000, would appear in the equation as a divisor, and in the final result would greatly reduce whatever errors existed in the observations employed.

Repeat Exercise 6 night after night, and note whether the stars come back to the same position at the same hour and minute every night.

11. ***The plumb-line apparatus.*** – This experiment, and many others, may be conveniently and accurately made with no other apparatus than a plumb line, and a device for sighting past it. In Figs. 6 and 7 there is shown a simple form of such apparatus, consisting essentially of a board which rests in a horizontal position upon the points of three screws that pass through it.

This board carries a small box, to one side of which is nailed in vertical position another board 5 or 6 feet long to carry the plumb line. This consists

FIG. 6. *FIG. 7.*

The plumb-line apparatus.

of a wire or fish line with any heavy weight— e. g., a brick or flatiron— tied to its lower end and immersed in a vessel of water placed inside the box, so as to check any swinging motion of the weight. In the cover of the box is a small hole through which the wire passes, and by turning the screws in the baseboard the apparatus may be readily leveled, so that the wire shall swing freely in the center of the hole without touching the cover of the box. Guy wires, shown in the figure, are applied so as to stiffen the whole apparatus. A board with a screw eye at each end may be pivoted to the upright, as in Fig. 6, for measuring altitudes; or to the box, as in Fig. 7, for observing the time at which a star in its diurnal motion passes through the plane determined by the plumb line and the center of the screw eye through which the observer looks.

The whole apparatus may be constructed by any person of ordinary mechanical skill at a very small cost, and it or something equivalent should be provided for every class beginning observational astronomy. To use the apparatus for the experiment of § 10, it should be leveled, and the board with the screw eyes, attached as in Fig. 7, should be turned until the observer, looking through the screw eye, sees Polaris exactly behind the wire. Use a bicycle lamp to illumine the wire by night. The apparatus is now adjusted, and the observer has only to wait for the stars which he desires to observe, and to note by his watch the time at which they pass behind the wire. It will be seen that the wire takes the place of the vertical edge of the building, and that the board with the screw eyes is introduced solely to keep the observer in the right place relative to the wire.

12. ***A sidereal clock.***— Clocks are sometimes so made and regulated that they show always the same hour and minute when the stars come back to the same place, and such a timepiece is called a sidereal clock— i. e., a star-time clock. Would such a clock gain or lose in comparison with an ordinary watch? Could an ordinary watch be turned into a sidereal watch by moving the regulator?

*FIG. 8.— **Photographing the circumpolar stars.**— BARNARD.*

13. ***Photographing the stars.***— EXERCISE 7.— For any student who uses a camera. Upon some clear and moonless night point the camera, properly focused, at Polaris, and expose a plate for three or four hours. Upon developing the plate you should find a series of circular trails such as are shown in Fig. 8, only longer. Each one of these is produced by a star moving slowly over the plate, in consequence of its changing position in the sky. The center indicated by these curved trails is called the pole of the heavens. It is

that part of the sky toward which is pointed the axis about which the earth rotates, and the motion of the stars around the center is only an apparent motion due to the rotation of the earth which daily carries the observer and his camera around this axis while the stars stand still, just as trees and fences and telegraph poles stand still, although to the passenger upon a railway train they appear to be in rapid motion. So far as simple observations are concerned, there is no method by which the pupil can tell for himself that the motion of the stars is an apparent rather than a real one, and, following the custom of astronomers, we shall habitually speak as if it were a real movement of the stars. How long was the plate exposed in photographing Fig. 8?

14. ***Finding the stars.***— On Plate I, opposite page 124, the pole of the heavens is at the center of the map, near Polaris, and the heavy trail near the center of Fig. 8 is made by Polaris. See if you can identify from the map any of the stars whose trails show in the photograph. The brighter the star the bolder and heavier its trail.

Find from the map and locate in the sky the two bright stars Capella and Vega, which are on opposite sides of Polaris and nearly equidistant from it. Do these stars share in the motion around the pole? Are they visible on every clear night, and all night?

Observe other bright stars farther from Polaris than are Vega and Capella and note their movement. Do they move like the sun and moon? Do they rise and set?

In what part of the sky do the stars move most rapidly, near the pole or far from it?

How long does it take the fastest moving stars to make the circuit of the sky and come back to the same place? How long does it take the slow stars?

15. ***Rising and setting of the stars.***— A study of the sky along the lines indicated in these questions will show that there is a considerable part of it surrounding the pole whose stars are visible on every clear night. The same star is sometimes high in the sky, sometimes low, sometimes to the east of the pole and at other times west of it, but is always above the horizon. Such stars are said to be circumpolar. A little farther from the pole each star, when at the lowest point of its circular path, dips for a time below the horizon and is lost to view, and the farther it is away from the pole the longer does it remain invisible, until, in the case of stars 90° away from the pole, we find them hidden below the horizon for twelve hours out of every twenty-four (see Fig. 9). The sun is such a star, and in its rising and setting acts precisely as does every other star at a similar distance from the pole— only, as we shall find later, each star keeps always at (nearly) the same distance from the pole, while the sun in the course of a year changes its distance from the pole very greatly, and thus changes the amount of time it spends above and below the horizon, producing in this way the long days of summer and the short ones of winter.

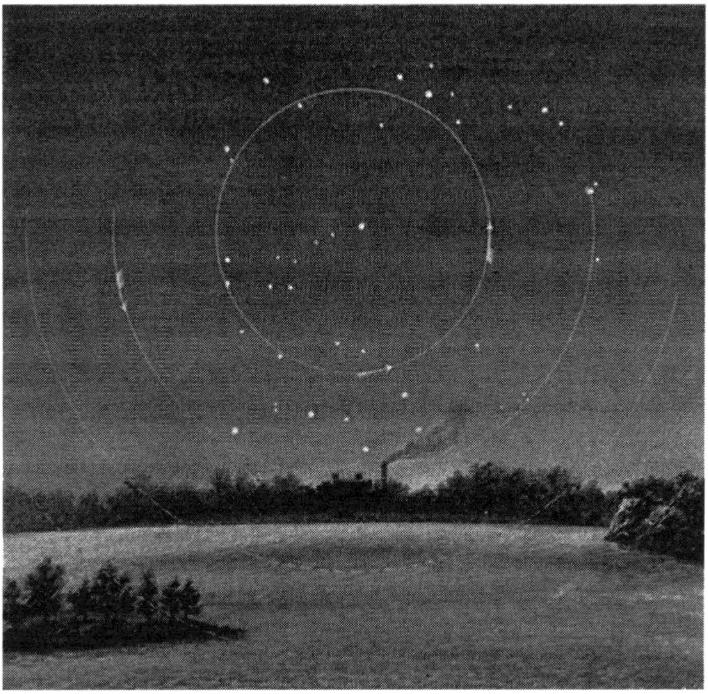

FIG. 9.— Diurnal motion of the northern constellations.

How much time do stars which are more than 90° from the pole spend above the horizon?

We say in common speech that the sun rises in the east, but this is strictly true only at the time when it is 90° distant from the pole— i. e., in March and September. At other seasons it rises north or south of east according as its distance from the pole is less or greater than 90°, and the same is true for the stars.

16. ***The geography of the sky.***— Find from a map the latitude and longitude of your schoolhouse. Find on the map the place whose latitude is 39° and longitude 77° west of the meridian of Greenwich. Is there any other place in the world which has the same latitude and longitude as your schoolhouse?

The places of the stars in the sky are located in exactly the manner which is illustrated by these geographical questions, only different names are used. Instead of latitude the astronomer says *declination*, in place of longitude he says *right ascension*, in place of meridian he says *hour circle*, but he means by these new names the same ideas that the geographer expresses by the old ones.

Imagine the earth swollen up until it fills the whole sky; the earth's equator would meet the sky along a line (a great circle) everywhere 90° distant from the pole, and this line is called the *celestial equator*. Trace its position along the middle of the map opposite page 190 and notice near what stars it runs. Every meridian of the swollen earth would touch the sky along an hour circle– i. e., a great circle passing through the pole and therefore perpendicular to the equator. Note that in the map one of these hour circles is marked 0. It plays the same part in measuring right ascensions as does the meridian of Greenwich in measuring longitudes; it is the beginning, from which they are reckoned. Note also, at the extreme left end of the map, the four bright stars in the form of a square, one side of which is parallel and close to the hour circle, which is marked 0. This is familiarly called the Great Square in Pegasus, and may be found high up in the southern sky whenever the Big Dipper lies below the pole. Why can it not be seen when Ursa Major is above the pole?

Astronomers use the right ascensions of the stars not only to tell in what part of the sky the star is placed, but also in time reckonings, to regulate their sidereal clocks, and with regard to this use they find it convenient to express right ascension not in degrees but in hours, 24 of which fill up the circuit of the sky and each of which is equal to 15° of arc, 24 × 15 = 360. The right ascension of Capella is 5h. 9m. = 77.2°, but the student should accustom himself to using it in hours and minutes as given and not to change it into degrees. He should also note that some stars lie on the side of the celestial equator toward Polaris, and others are on the opposite side, so that the astronomer has to distinguish between north declinations and south declinations, just as the geographer distinguishes between north latitudes and south latitudes. This is done by the use of the + and - signs, a + denoting that the star lies north of the celestial equator, i. e., toward Polaris.

Find on Plate II, opposite page 190, the Pleiades (Plēadēs), R. A. = 3h. 42m., Dec. = +23.8°. Why do they not show on Plate I, opposite page 124? In what direction are they from Polaris? This is one of the finest star clusters in the sky, but it needs a telescope to bring out its richness. See how many stars you can count in it with the naked eye, and afterward examine it with an opera glass. Compare what you see with Fig. 10. Find Antares, R. A. = 16h. 23m. Dec. = -26.2°. How far is it, in degrees, from the pole? Is it visible in your sky? If so, what is its color?

Find the R. A. and Dec. of α Ursæ Majoris; of β Ursæ Majoris; of Polaris. Find the Northern Crown, *Corona Borealis*, R. A. = 15h. 30m., Dec. = +27.0°; the Beehive, *Præsepe*, R. A. = 8h. 33m., Dec. = +20.4°.

These should be looked up, not only on the map, but also in the sky.

FIG. 10.— From a photograph of the Pleiades.

17. ***Reference lines and circles.***— As the stars move across the sky in their diurnal motion, they carry the framework of hour circles and equator with them, so that the right ascension and declination of each star remain unchanged by this motion, just as longitudes and latitudes remain unchanged by the earth's rotation. They are the same when a star is rising and when it is setting; when it is above the pole and when it is below it. During each day the hour circle of every star in the heavens passes overhead, and at the moment when any particular hour circle is exactly overhead all the stars which lie upon it are said to be "on the meridian"— i. e., at that particular moment they stand directly over the observer's geographical meridian and upon the corresponding celestial meridian.

An eye placed at the center of the earth and capable of looking through its solid substance would see your geographical meridian against the background of the sky exactly covering your celestial meridian and passing from one pole through your zenith to the other pole. In Fig. 11 the inner circle represents the terrestrial meridian of a certain place, *O*, as seen from the center of the earth, *C*, and the outer circle represents the celestial meridian of *O* as seen from *C*, only we must imagine, what can not be shown on the figure, that the outer circle is so large that the inner one shrinks to a mere point in comparison with it. If *CP* represents the direction in which the earth's axis passes through the center, then *CE* at right angles to it must be the direction of the equator which we suppose to be turned edgewise toward us; and if *C O* is the direction of some particular point on the earth's surface, then *Z* directly overhead is called the *zenith* of that point, upon the celestial sphere. The line *C H* represents a direction parallel to the horizon plane at *O*, and *H C P* is the angle which the axis of the earth makes with this horizon plane. The arc *O E* measures the latitude of *O*, and the arc *Z E* measures the declination of *Z*, and since by elementary geometry each of these arcs contains the same number of degrees as the angle *E C Z*, we have the

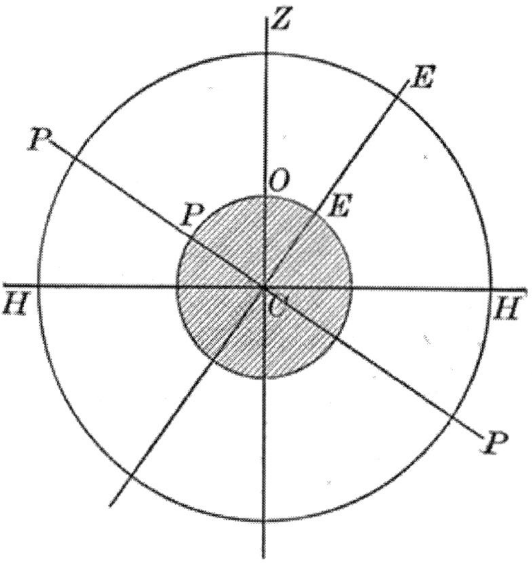

*FIG. 11.— **Reference lines and circles.***

Theorem.— The latitude of any place is equal to the declination of its zenith.

Corollary.— Any star whose declination is equal to your latitude will once in each day pass through your zenith.

18. ***Latitude.*** — From the construction of the figure

$$\angle ECZ + \angle ZCP = 90^\circ$$

$$\angle HCP + \angle ZCP = 90^\circ$$

from which we find by subtraction and transposition

$\angle ECZ = \angle HCP$

and this gives the further

Theorem.— The latitude of any place is equal to the elevation of the pole above its horizon plane.

An observer who travels north or south over the earth changes his latitude, and therefore changes the angle between his horizon plane and the axis of the earth. What effect will this have upon the position of stars in his sky? If you were to go to the earth's equator, in what part of the sky would you look for Polaris? Can Polaris be seen from Australia? From South America? If you were to go from Minnesota to Texas, in what respect would the appearance of stars in the northern sky be changed? How would the appearance of stars in the southern sky be changed?

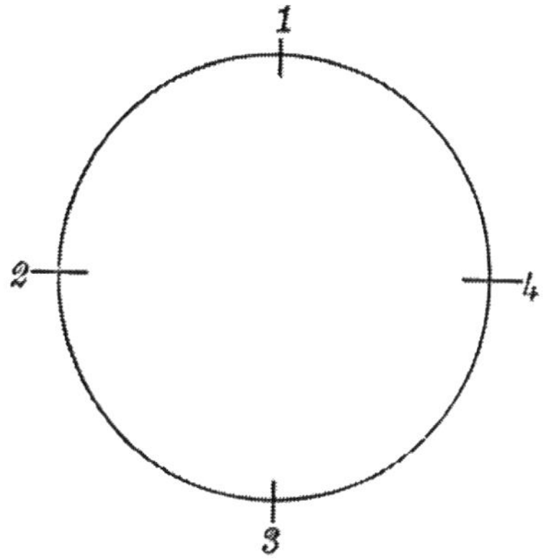

FIG. 12 — Diurnal path of Polaris.

EXERCISE 8.– Determine your latitude by taking the altitude of Polaris when it is at some one of the four points of its diurnal path, shown in Fig. 12. When it is at *1* it is said to be at upper culmination, and the star ζ Ursæ Majoris in the handle of the Big Dipper will be directly below it. When at *2* it is at western elongation, and the star Castor is near the meridian. When it is at *3* it is at lower culmination, and the star Spica is on the meridian. When it is at *4* it is at eastern elongation, and Altair is near the meridian. All of these stars are conspicuous ones, which the student should find upon the map and learn to recognize in the sky. The altitude observed at either *2* or *4* may be considered equal to the latitude of the place, but the altitude observed when Polaris is at the positions marked *1* and *3* must be corrected for the star's distance from the pole, which may be assumed equal to 1.3°.

The plumb-line apparatus described at page 12 is shown in Fig. 6 slightly modified, so as to adapt it to measuring the altitudes of stars. Note that the board with the screw eye at one end has been transferred from the box to the vertical standard, and has a screw eye at each end. When the apparatus has been properly leveled, so that the plumb line hangs at the middle of the hole in the box cover, the board is to be pointed at the star by sighting through the centers of the two screw eyes, and a pencil line is to be ruled along its edge upon the face of the vertical standard. After this has been done turn the apparatus halfway around so that what was the north side now points south, level it again and revolve the board about the screw which holds it to the vertical standard, until the screw eyes again point to the star. Rule another line along the same edge of the board as before and with a protractor measure the angle between these lines. Use a bicycle lamp if you need artificial light for your work. The student who has studied plane geometry should be able to prove that one half of the angle between these lines is equal to the altitude of the star.

After you have determined your latitude from Polaris, compare the result with your position as shown upon the best map available. With a little practice and considerable care the latitude may be thus determined within one tenth of a degree, which is equivalent to about 7 miles. If you go 10 miles north or south from your first station you should find the pole higher up or lower down in the sky by an amount which can be measured with your apparatus.

19. ***The meridian line.***— To establish a true north and south line upon the ground, use the apparatus as described at page 13, and when Polaris is at upper or lower culmination drive into the ground two stakes in line with the star and the plumb line. Such a meridian line is of great convenience in observing the stars and should be laid out and permanently marked in some convenient open space from which, if possible, all parts of the sky are visible. June and November are convenient months for this exercise, since Polaris then comes to culmination early in the evening.

20. ***Time.***— What is *the time* at which school begins in the morning? What do you mean by "*the time*"?

The sidereal time at any moment is the right ascension of the hour circle which at that moment coincides with the meridian. When the hour circle passing through Sirius coincides with the meridian, the sidereal time is 6h. 40m., since that is the right ascension of Sirius, and in astronomical language Sirius is "*on the meridian*" at 6h. 40m. sidereal time. As may be seen from the map, this 6h. 40m. is the right ascension of Sirius, and if a clock be set to indicate 6h. 40m. when Sirius crosses the meridian, it will show sidereal time. If the clock is properly regulated, every other star in the heavens will come to the meridian at the moment when the time shown by the clock is equal to the right ascension of the star. A clock properly regulated for this purpose will gain about four minutes per day in comparison with ordinary clocks, and when so regulated it is called a sidereal clock. The student should be provided with such a clock for his future work, but one such clock will serve for several persons, and a nutmeg clock or a watch of the cheapest kind is quite sufficient.

The Harvard College Observatory, Cambridge, Mass.

EXERCISE 9.– Set such a clock to sidereal time by means of the transit of a star over your meridian. For this experiment it is presupposed that a meridian line has been marked out on the ground as in § 19, and the simplest mode of performing the experiment required is for the observer, having chosen a suitable star in the southern part of the sky, to place his eye accurately over the northern end of the meridian line and to estimate as nearly as possible the beginning and end of the period during which the star appears to stand exactly above the southern end of the line. The middle of this period may be taken as the time at which the star crossed the meridian and at this moment the sidereal time is equal to the right ascension of the star. The difference between this right ascension and the observed middle instant is the error of the clock or the amount by which its hands must be set back or forward in order to indicate true sidereal time.

A more accurate mode of performing the experiment consists in using the plumb-line apparatus carefully adjusted, as in Fig. 7, so that the line joining the wire to the center of the screw eye shall be parallel to the meridian line. Observe the time by the clock at which the star disappears behind the wire as seen through the center of the screw eye. If the star is too high up in the sky for convenient observation, place a mirror, face up, just north of the screw eye and observe star, wire and screw eye by reflection in it.

The numerical right ascension of the observed star is needed for this experiment, and it may be measured from the star map, but it will usually be best to observe one of the stars of the table at the end of the book, and to obtain its right ascension as follows: The table gives the right ascension and declination of each star as they were at the beginning of the year 1900, but on account of the precession (see Chapter V), these numbers all change slowly with the lapse of time, and on the average the right ascension of each star of the table must be increased by one twentieth of a minute for each year after 1900– i. e., in 1910 the right ascension of the first star of the table will be 0h. 38.6m. + (10/20)m. = 0h. 39.1m. The declinations also change slightly, but as they are only intended to help in finding the star on the star maps, their change may be ignored.

Having set the clock approximately to sidereal time, observe one or two more stars in the same way as above. The difference between the observed time and the right ascension, if any is found, is the "correction" of the clock. This correction ought not to exceed a minute if due care has been taken in the several operations prescribed. The relation of the clock to the right ascension of the stars is expressed in the following equation, with which the student should become thoroughly familiar:

$$A = T \pm U$$

T stands for the time by the clock at which the star crossed the meridian. A is the right ascension of the star, and U is the correction of the clock. Use the + sign in the equation whenever the clock is too slow, and the - sign when it is too fast. U may be found from this equation when A and T are given, or A may be found when T and U are given. It is in this way that astronomers measure the right ascensions of the stars and planets.

Determine U from each star you have observed, and note how the several results agree one with another.

21. ***Definitions.***— To define a thing or an idea is to give a description sufficient to identify it and distinguish it from every other possible thing or idea. If a definition does not come up to this standard it is insufficient. Anything beyond this requirement is certainly useless and probably mischievous.

Let the student define the following geographical terms, and let him also criticise the definitions offered by his fellow-students: Equator, poles, meridian, latitude, longitude, north, south, east, west.

Compare the following astronomical definitions with your geographical definitions, and criticise them in the same way. If you are not able to improve upon them, commit them to memory:

The Poles of the heavens are those points in the sky toward which the earth's axis points. How many are there? The one near Polaris is called the north pole.

The Celestial Equator is a great circle of the sky distant 90° from the poles.

The Zenith is that point of the sky, overhead, toward which a plumb line points. Why is the word overhead placed in the definition? Is there more than one zenith?

The Horizon is a great circle of the sky 90° distant from the zenith.

An Hour Circle is any great circle of the sky which passes through the poles. Every star has its own hour circle.

The Meridian is that hour circle which passes through the zenith.

A Vertical Circle is any great circle that passes through the zenith. Is the meridian a vertical circle?

The Declination of a star is its angular distance north or south of the celestial equator.

The Right Ascension of a star is the angle included between its hour circle and the hour circle of a certain point on the equator which is called the *Vernal Equinox*. From spherical geometry we learn that this angle is to be measured either at the pole where the two hour circles intersect, as is done in the star map opposite page 124, or along the equator, as is done in the map opposite page 190. Right ascension is always measured from the vernal equinox in the direction opposite to that in which the stars appear to travel in their diurnal motion— i. e., from west toward east.

The Altitude of a star is its angular distance above the horizon.

The Azimuth of a star is the angle between the meridian and the vertical circle passing through the star. A star due south has an azimuth of 0°. Due west, 90°. Due north, 180°. Due east, 270°.

What is the azimuth of Polaris in degrees?

What is the azimuth of the sun at sunrise? At sunset? At noon? Are these azimuths the same on different days?

The Hour Angle of a star is the angle between its hour circle and the meridian. It is measured from the meridian in the direction in which the stars appear to travel in their diurnal motion— i. e., from east toward west.

What is the hour angle of the sun at noon? What is the hour angle of Polaris when it is at the lowest point in its daily motion?

22. ***Exercises.***— The student must not be satisfied with merely learning these definitions. He must learn to see these points and lines in his mind as if they were visibly painted upon the sky. To this end it will help him to note that the poles, the zenith, the meridian, the horizon, and the equator seem to stand still in the sky, always in the same place with respect to the observer, while the hour circles and the vernal equinox move with the stars and keep the same place among them. Does the apparent motion of a star change its declination or right ascension? What is the hour angle of the sun when it has the greatest altitude? Will your answer to the preceding question be true for a star? What is the altitude of the sun after sunset? In what direction is the north pole from the zenith? From the vernal equinox? Where are the points in which the meridian and equator respectively intersect the horizon?

CHAPTER III.
FIXED AND WANDERING STARS

23. ***Star maps.***— Select from the map some conspicuous constellation that will be conveniently placed for observation in the evening, and make on a large scale a copy of all the stars of the constellation that are shown upon the map. At night compare this copy with the sky, and mark in upon your paper all the stars of the constellation which are not already there. Both the original drawing and the additions made to it by night should be carefully done, and for the latter purpose what is called the method of allineations may be used with advantage— i. e., the new star is in line with two already on the drawing and is midway between them, or it makes an equilateral triangle with two others, or a square with three others, etc.

A series of maps of the more prominent constellations, such as Ursa Major, Cassiopea, Pegasus, Taurus, Orion, Gemini, Canis Major, Leo, Corvus, Bootes, Virgo, Hercules, Lyra, Aquila, Scorpius, should be constructed in this manner upon a uniform scale and preserved as a part of the student's work. Let the magnitude of the stars be represented on the maps as accurately as may be, and note the peculiarity of color which some stars present. For the most part their color is a very pale yellow, but occasionally one may be found of a decidedly ruddy hue— e. g., Aldebaran or Antares. Such a star map, not quite complete, is shown in Fig. 13.

So, too, a sharp eye may detect that some stars do not remain always of the same magnitude, but change their brightness from night to night, and this not on account of cloud or mist in the atmosphere, but from something in the star itself. Algol is one of the most conspicuous of these *variable stars*, as they are called.

FIG. 13.— Star map of the region about Orion.

24. ***The moon's motion among the stars.***— Whenever the moon is visible note its position among the stars by allineations, and plot it on the key map opposite page 190. Keep a record of the day and hour corresponding to each such observation. You will find, if the work is correctly done, that the positions of the moon all fall near the curved line shown on the map. This line is called the ecliptic.

After several such observations have been made and plotted, find by measurement from the map how many degrees per day the moon moves. How long would it require to make the circuit of the heavens and come back to the starting point?

On each night when you observe the moon, make on a separate piece of paper a drawing of it about 10 centimeters in diameter and show in the drawing every feature of the moon's face which you can see— e. g., the shape of the illuminated surface (phase); the direction among the stars of the line joining the horns; any spots which you can see upon the moon's face, etc. An opera glass will prove of great assistance in this work.

Use your drawings and the positions of the moon plotted upon the map to answer the following questions: Does the direction of the line joining the horns have any special relation to the ecliptic? Does the amount of illuminated surface of the moon have any relation to the moon's angular distance from the sun? Does it have any relation to the time at which the moon sets? Do the spots on the moon when visible remain always in the same place? Do they come and go? Do they change their position with relation to each other? Can you determine from these spots that the moon rotates about an axis, as the earth does? In what direction does its axis point? How long does it take to make one revolution about the axis? Is there any day and night upon the moon?

Each of these questions can be correctly answered from the student's own observations without recourse to any book.

25. ***The sun and its motion.***— Examine the face of the sun through a smoked glass to see if there is anything there that you can sketch.

By day as well as by night the sky is studded with stars, only they can not be seen by day on account of the overwhelming glare of sunlight, but the position of the sun among the stars may be found quite as accurately as was that of the moon, by observing from day to day its right ascension and declination, and this should be practiced at noon on clear days by different members of the class.

EXERCISE 10.— The right ascension of the sun may be found by observing with the sidereal clock the time of its transit over the meridian. Use the equation in § 20, and substitute in place of U the value of the clock correction found from observations of stars on a preceding or following night. If the clock gains or loses *with respect to sidereal time*, take this into account in the value of *U*.

EXERCISE 11.– To determine the sun's declination, measure its altitude at the time it crosses the meridian. Use either the method of Exercise 4, or that used with Polaris in Exercise 8. The student should be able to show from Fig. 11 that the declination is equal to the sum of the altitude and the latitude of the place diminished by 90°, or in an equation

Declination = Altitude + Latitude - 90°.

If the declination as found from this equation is a negative number it indicates that the sun is on the south side of the equator.

The right ascension and declination of the sun as observed on each day should be plotted on the map and the date, written opposite it. If the work has been correctly done, the plotted points should fall upon the curved line (ecliptic) which runs lengthwise of the map. This line, in fact, represents the sun's path among the stars.

Note that the hours of right ascension increase from 0 up to 24, while the numbers on the clock dial go only from 0 to 12, and then repeat 0 to 12 again during the same day. When the sidereal time is 13 hours, 14 hours, etc., the clock will indicate 1 hour, 2 hours, etc., and 12 hours must then be added to the time shown on the dial.

If observations of the sun's right ascension and declination are made in the latter part of either March or September the student will find that the sun crosses the equator at these times, and he should determine from his observations, as accurately as possible, the date and hour of this crossing and the point on the equator at which the sun crosses it. These points are called the equinoxes, Vernal Equinox and Autumnal Equinox for the spring and autumn crossings respectively, and the student will recall that the vernal equinox is the point from which right ascensions are measured. Its position among the stars is found by astronomers from observations like those above described, only made with much more elaborate apparatus.

Similar observations made in June and December show that the sun's midday altitude is about 47° greater in summer than in winter. They show also that the sun is as far north of the equator in June as he is south of it in December, from which it is easily inferred that his path, the ecliptic, is inclined to the equator at an angle of 23°.5, one half of 47°. This angle is called the obliquity of the ecliptic. The student may recall that in the geographies the torrid zone is said to extend 23°.5 on either side of the earth's equator. Is there any connection between these limits and the obliquity of the ecliptic? Would it be correct to define the torrid zone as that part of the earth's surface within which the sun may at some season of the year pass through the zenith?

EXERCISE 12.– After a half dozen observations of the sun have been plotted upon the map, find by measurement the rate, in degrees per day, at which the sun moves along the ecliptic. How many days will be required for it to move completely around the ecliptic from vernal equinox back to vernal equinox again? Accurate observations with the elaborate apparatus used by professional astronomers show that this period, which is called a *tropical year*, is 365 days 5 hours 48 minutes 46 seconds. Is this the same as the ordinary year of our calendars?

26. ***The planets.***– Any one who has watched the sky and who has made the drawings prescribed in this chapter can hardly fail to have found in the course of his observations some bright stars not set down on the printed star maps, and to have found also that these stars do not remain fixed in position among their fellows, but wander about from one constellation to another. Observe the motion of one of these planets from night to night and plot its positions on the star map, precisely as was done for the moon. What kind of path does it follow?

Both the ancient Greeks and the modern Germans have called these bodies wandering stars, and in English we name them planets, which is simply the Greek word for wanderer, bent to our use. Besides the sun and moon there are in the heavens five planets easily visible to the naked eye and, as we shall see later, a great number of smaller ones visible only in the telescope. More than 2,000 years ago astronomers began observing the motion of sun, moon, and planets among the stars, and endeavored to account for these motions by the theory that each wandering star moved in an orbit about the earth. Classical and mediæval literature are permeated with this idea, which was displaced only after a long struggle begun by Copernicus (1543 A. D.), who taught that the moon alone of these bodies revolves about the earth, while the earth and the other planets revolve around the sun. The ecliptic is the intersection of the plane of the earth's orbit with the sky, and the sun appears to move along the ecliptic because, as the earth moves around its orbit, the sun is always seen projected against the opposite side of it. The moon and planets all appear to move near the ecliptic because the planes of their orbits nearly coincide with the plane of the earth's orbit, and a narrow strip on either side of the ecliptic, following its course completely around the sky, is called the *zodiac*, a word which may be regarded as the name of a narrow street (16° wide) within which all the wanderings of the visible planets are confined and outside of which they never venture. Indeed, Mars is the only planet which ever approaches the edge of the street, the others traveling near the middle of the road.

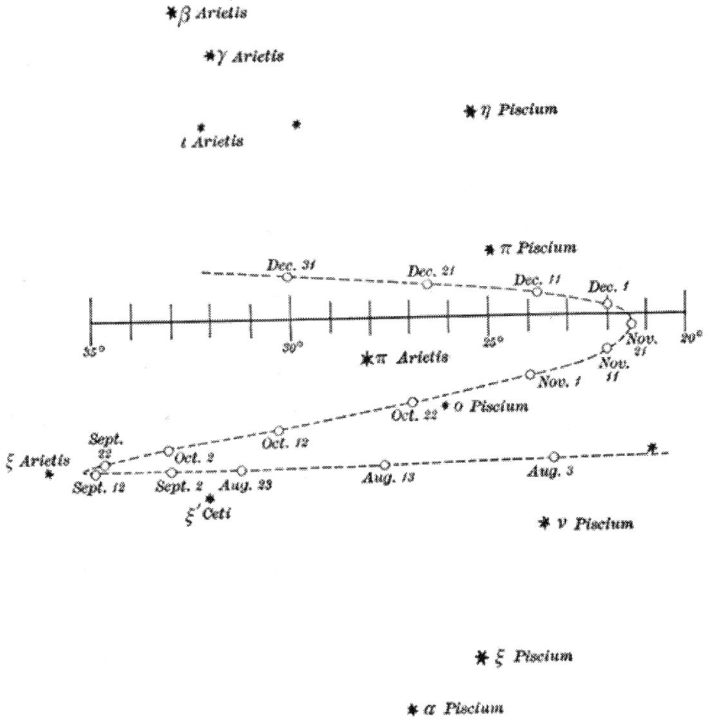

*FIG. 14.— **The apparent motion of a planet.***

27. ***A typical case of planetary motion.***— The Copernican theory, enormously extended and developed through the Newtonian law of gravitation (see Chapter IV), has completely supplanted the older Ptolemaic doctrine, and an illustration of the simple manner in which it accounts for the apparently complicated motions of a planet among the stars is found in Figs. 14 and 15, the first of which represents the apparent motion of the planet Mars through the constellations Aries and Pisces during the latter part of the year 1894, while the second shows the true motions of Mars and the earth in their orbits about the sun during the same period. The straight line in Fig. 14, with cross ruling upon it, is a part of the ecliptic, and the numbers placed opposite it represent the distance, in degrees, from the vernal equinox. In Fig. 15 the straight line represents the direction from the sun toward the vernal equinox, and the angle which this line makes with the line joining earth and sun is called the earth's longitude. The imaginary line joining the earth and sun is called the earth's radius vector, and the pupil should note that the longitude and length of the radius vector taken together show the direction and distance of the earth from the sun— i. e., they fix the relative positions of the two bodies. The same is nearly true for Mars and would be wholly true if the orbit of Mars lay in the same plane with that of the earth. How

does Fig. 14 show that the orbit of Mars does not lie exactly in the same plane with the orbit of the earth?

EXERCISE 13.– Find from Fig. 15 what ought to have been the apparent course of Mars among the stars during the period shown in the two figures, and compare what you find with Fig. 14. The apparent position of Mars among the stars is merely its direction from the earth, and this direction is represented in Fig. 14 by the distance of the planet from the ecliptic and by its longitude.

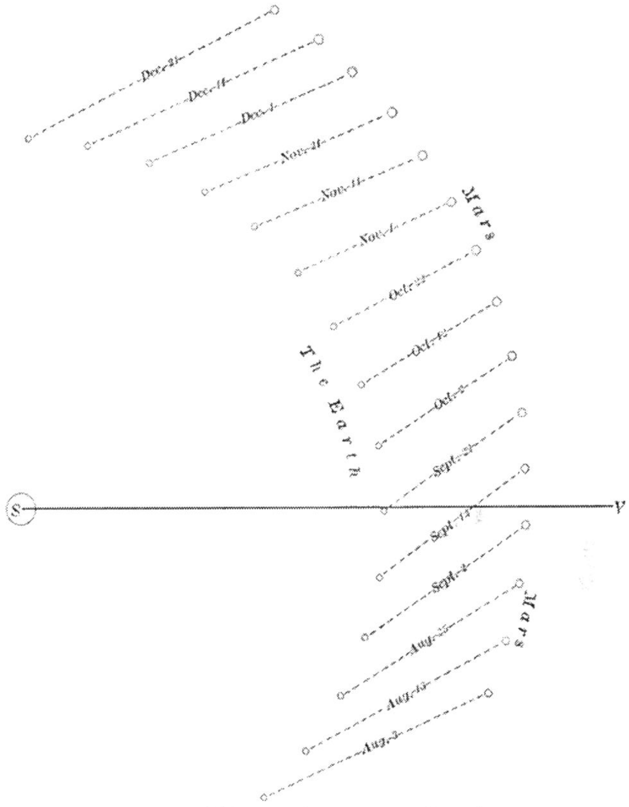

*FIG. 15.— **The real motion of a planet.***

The longitude of Mars for each date can be found from Fig. 15 by measuring the angle between the straight line *S V* and the line drawn from the earth to Mars. Thus for October 12th we may find with the protractor that the angle between the line *S V* and the line joining the earth to Mars is a little more than 30°, and in Fig. 14 the position of Mars for this date is shown nearly opposite the cross line corresponding to 30° on the ecliptic. Just how far below the ecliptic this position of Mars should fall can not be told from Fig. 15, which from necessity is constructed as if the orbits of Mars and the earth lay in the same plane, and Mars in this case would always appear to stand exactly on the ecliptic and to oscillate back and forth as shown in Fig. 14, but without the up-and-down motion there shown. In this way plot in Fig. 14 the longitudes of Mars as seen from the earth for other dates and observe how the forward motion of the two planets in their orbits accounts for the apparently capricious motion of Mars to and fro among the stars.

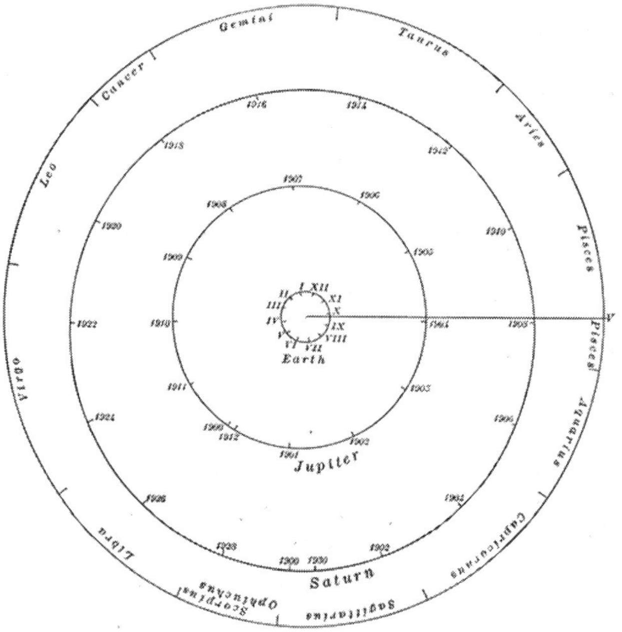

*FIG. 16.— **The orbits of Jupiter and Saturn.***

28. ***The orbits of the planets.*** — Each planet, great or small, moves in its own appropriate orbit about the sun, and the exact determination of these orbits, their sizes, shapes, positions, etc., has been one of the great problems of astronomy for more than 2,000 years, in which successive generations of astronomers have striven to push to a still higher degree of accuracy the knowledge attained by their predecessors. Without attempting to enter into the details of this problem we may say, generally, that every planet moves in a plane passing through the sun, and for the six planets visible to the naked eye these planes nearly coincide, so that the six orbits may all be shown without much error as lying in the flat surface of one map. It is, however, more convenient to use two maps, such as Figs. 16 and 17, one of which shows the group of planets, Mercury, Venus, the earth, and Mars, which are near the sun, and on this account are sometimes called the inner planets, while the other shows the more distant planets, Jupiter and Saturn, together with

the earth, whose orbit is thus made to serve as a connecting link between the two diagrams. These diagrams are accurately drawn to scale, and are intended to be used by the student for accurate measurement in connection with the exercises and problems which follow.

In addition to the six planets shown in the figures the solar system contains two large planets and several hundred small ones, for the most part invisible to the naked eye, which are omitted in order to avoid confusing the diagrams.

29. ***Jupiter and Saturn.*** – In Fig. 16 the sun at the center is encircled by the orbits of the three planets, and inclosing all of these is a circular border showing the directions from the sun of the constellations which lie along the zodiac. The student must note carefully that it is only the directions of these constellations that are correctly shown, and that in order to show them at all they have been placed very much too close to the sun. The cross lines extending from the orbit of the earth toward the sun with Roman numerals opposite them show the positions of the earth in its orbit on the first day of January (*I*), first day of February (*II*), etc., and the similar lines attached to the orbits of Jupiter and Saturn with Arabic numerals show the positions of those planets on the first day of January of each year indicated, so that the figure serves to show not only the orbits of the planets, but their actual positions in their orbits for something more than the first decade of the twentieth century.

The line drawn from the sun toward the right of the figure shows the direction to the vernal equinox. It forms one side of the angle which measures a planet's longitude.

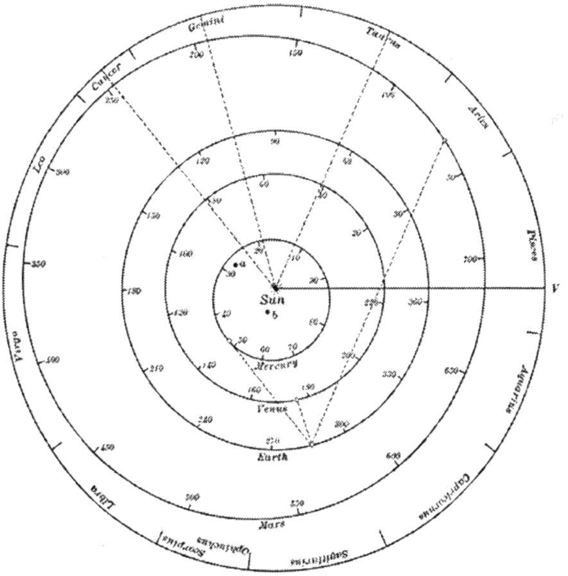

*FIG. 17.– **The orbits of the inner planets.***

EXERCISE 14.– Measure with your protractor the longitude of the earth on January 1st. Is this longitude the same in all years? Measure the longitude of Jupiter on January 1, 1900; on July 1, 1900; on September 25, 1906.

Draw neatly on the map a pencil line connecting the position of the earth for January 1, 1900, with the position of Jupiter for the same date, and produce the line beyond Jupiter until it meets the circle of the constellations. This line represents the direction of Jupiter from the earth, and points toward the constellation in which the planet appears at that date. But this representation of the place of Jupiter in the sky is not a very accurate one, since on the scale of the diagram the stars are in fact more than 100,000 times as far off as they are shown in the figure, and the pencil mark does not meet the line of constellations at the same intersection it would have if this line were pushed back to its true position. To remedy this defect we must draw another line from the sun parallel to the one first drawn, and its intersection with the constellations will give very approximately the true position of Jupiter in the sky.

EXERCISE 15.– Find the present positions of Jupiter and Saturn, and look them up in the sky by means of your star maps. The planets will appear in the indicated constellations as very bright stars not shown on the map.

Which of the planets, Jupiter and Saturn, changes its direction from the sun more rapidly? Which travels the greater number of miles per day? When will Jupiter and Saturn be in the same constellation? Does the earth move faster or slower than Jupiter?

The distance of Jupiter or Saturn from the earth at any time may be readily obtained from the figure. Thus, by direct measurement with the millimeter scale we find for January 1, 1900, the distance of Jupiter from the earth is 6.1 times the distance of the sun from the earth, and this may be turned into miles by multiplying it by 93,000,000, which is approximately the distance of the sun from the earth. For most purposes it is quite as well to dispense with this multiplication and call the distance 6.1 astronomical units, remembering that the astronomical unit is the distance of the sun from the earth.

EXERCISE 16.– What is Jupiter's distance from the earth at its nearest approach? What is the greatest distance it ever attains? Is Jupiter's least distance from the earth greater or less than its least distance from Saturn?

On what day in the year 1906 will the earth be on line between Jupiter and the sun? On this day Jupiter is said to be in *opposition*– i. e., the planet and the sun are on opposite sides of the earth, and Jupiter then comes to the meridian of any and every place at midnight. When the sun is between the earth and Jupiter (at what date in 1906?) the planet is said to be in *conjunction* with the sun, and of course passes the meridian with the sun at noon. Can you determine from the figure the time at which Jupiter comes to the meridian at other dates than opposition and conjunction? Can you determine when it is visible in the evening hours? Tell from the figure what constellation is on the meridian at midnight on January 1st. Will it be the same constellation in every year?

30. ***Mercury, Venus, and Mars.*** – Fig. 17, which represents the orbits of the inner planets, differs from Fig. 16 only in the method of fixing the positions of the planets in their orbits at any given date. The motion of these planets is so rapid, on account of their proximity to the sun, that it would not do to mark their positions as was done for Jupiter and Saturn, and with the exception of the earth they do not always return to the same place on the same day in each year. It is therefore necessary to adopt a slightly different method, as follows: The straight line extending from the sun toward the vernal equinox, *V*, is called the prime radius, and we know from past observations that the earth in its motion around the sun crosses this line on September 23d in each year, and to fix the earth's position for September 23d in the diagram we have only to take the point at which the prime radius intersects the earth's orbit. A month later, on October 23d, the earth will no longer be at this point, but will have moved on along its orbit to the point marked 30 (thirty days after September 23d). Sixty days after September 23d it will be at the point marked 60, etc., and for any date we have only to find the number of days intervening between it and the preceding September 23d, and this number will show at once the position of the earth in its orbit. Thus for the date July 4, 1900, we find

1900, July 4 - 1899, September 23 = 284 days,

and the little circle marked upon the earth's orbit between the numbers 270 and 300 shows the position of the earth on that date.

In what constellation was the sun on July 4, 1900? What zodiacal constellation came to the meridian at midnight on that date? What other constellations came to the meridian at the same time?

The positions of the other planets in their orbits are found in the same manner, save that they do not cross the prime radius on the same date in each year, and the times at which they do cross it must be taken from the following table:

Table of Epochs

A. D.	*Mercury.*	*Venus.*	*Earth.*	*Mars.*
Period	88.0 days.	224.7 days.	365.25 days.	687.1 days.
1900	Feb. 18th.	Jan. 11th.	Sept. 23d.	April 28th.
1901	Feb. 5th.	April 5th.	Sept. 23d.	...

A. D.	*Mercury.*	*Venus.*	*Earth.*	*Mars.*
1902	Jan. 23d.	June 29th.	Sept. 23d.	March 16th.
1903	April 8th.	Feb. 8th.	Sept. 23d.	...
1904	March 25th.	May 3d.	Sept. 23d.	Feb. 1st.
1905	March 12th.	July 26th.	Sept. 23d.	Dec. 19th.
1906	Feb. 27th.	March 8th.	Sept. 23d.	...
1907	Feb. 14th.	May 31st.	Sept. 23d.	Nov. 6th.
1908	Feb. 1st.	Jan. 11th.	Sept. 23d.	...
1909	Jan. 18th.	April 4th.	Sept. 23d.	Sept. 23d.
1910	Jan. 5th.	June 28th.	Sept. 23d.	...

The first line of figures in this table shows the number of days that each of these planets requires to make a complete revolution about the sun, and it appears from these numbers that Mercury makes about four revolutions in its orbit per year, and therefore crosses the prime radius four times in each year, while the other planets are decidedly slower in their movements. The following lines of the table show for each year the date at which each planet first crossed the prime radius in that year; the dates of subsequent crossings in any year can be found by adding once, twice, or three times the period to the given date, and the table may be extended to later years, if need be, by continuously adding multiples of the period. In the case of Mars it appears that there is only about one year out of two in which this planet crosses the prime radius.

After the date at which the planet crosses the prime radius has been determined its position for any required date is found exactly as in the case of the earth, and the constellation in which the planet will appear from the earth is found as explained above in connection with Jupiter and Saturn.

The broken lines in the figure represent the construction for finding the places in the sky occupied by Mercury, Venus, and Mars on July 4, 1900. Let the student make a similar construction and find the positions of these planets at the present time. Look them up in the sky and see if they are where your work puts them.

31. ***Exercises.***— The "evening star" is a term loosely applied to any planet which is visible in the western sky soon after sunset. It is easy to see that such a planet must be farther toward the east in the sky than is the sun, and in either Fig. 16 or Fig. 17 any planet which viewed from the position of the

earth lies to the left of the sun and not more than 50° away from it will be an evening star. If to the right of the sun it is a morning star, and may be seen in the eastern sky shortly before sunrise.

What planet is the evening star *now*? Is there more than one evening star at a time? What is the morning star now?

Do Mercury, Venus, or Mars ever appear in opposition? What is the maximum angular distance from the sun at which Venus can ever be seen? Why is Mercury a more difficult planet to see than Venus? In what month of the year does Mars come nearest to the earth? Will it always be brighter in this month than in any other? Which of all the planets comes nearest to the earth?

The earth always comes to the same longitude on the same day of each year. Why is not this true of the other planets?

The student should remember that in one respect Figs. 16 and 17 are not altogether correct representations, since they show the orbits as all lying in the same plane. If this were strictly true, every planet would move, like the sun, always along the ecliptic; but in fact all of the orbits are tilted a little out of the plane of the ecliptic and every planet in its motion deviates a little from the ecliptic, first to one side then to the other; but not even Mars, which is the most erratic in this respect, ever gets more than eight degrees away from the ecliptic, and for the most part all of them are much closer to the ecliptic than this limit.

CHAPTER IV.
CELESTIAL MECHANICS

32. ***The beginnings of celestial mechanics.***– From the earliest dawn of civilization, long before the beginnings of written history, the motions of sun and moon and planets among the stars from constellation to constellation had commanded the attention of thinking men, particularly of the class of priests. The religions of which they were the guardians and teachers stood in closest relations with the movements of the stars, and their own power and influence were increased by a knowledge of them.

Isaac Newton (1643-1727)

Out of these professional needs, as well as from a spirit of scientific research, there grew up and flourished for many centuries a study of the motions of the planets, simple and crude at first, because the observations that could then be made were at best but rough ones, but growing more accurate and more complex as the development of the mechanic arts put better and more precise instruments into the hands of astronomers and enabled them to observe with increasing accuracy the movements of these bodies. It was early seen that while for the most part the planets, including the sun and moon, traveled through the constellations from west to east, some of them sometimes reversed their motion and for a time traveled in the opposite way. This clearly can not be explained by the simple theory which had early been adopted that a planet moves always in the same direction around a circular orbit having the earth at its center, and so it was said to move around in a small circular orbit, called an epicycle, whose center was situated upon and moved along a circular orbit, called the deferent, within which the earth was placed, as is shown in Fig. 18, where the small circle is the epicycle, the large circle is the deferent, *P* is the planet, and *E* the earth. When this proved inadequate to account for the really complicated movements of the planets, another epicycle was put on top of the first one, and then another and another, until the supposed system became so complicated that Copernicus, a Polish astronomer, repudiated its fundamental theorem and taught that the motions of the planets take place in circles around the sun instead of about the earth, and that the earth itself is only one of the planets moving around the sun in its own appropriate orbit and itself largely responsible for the seemingly erratic movements of the other planets, since from day to day we see them and observe their positions from different points of view.

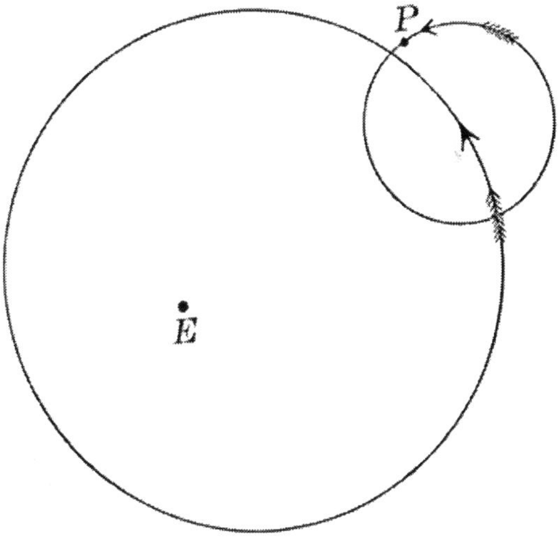

FIG. 18.— Epicycle and deferent.

33. ***Kepler's laws.***— Two generations later came Kepler with his three famous laws of planetary motion:

I. Every planet moves in an ellipse which has the sun at one of its foci.

II. The radius vector of each planet moves over equal areas in equal times.

III. The squares of the periodic times of the planets are proportional to the cubes of their mean distances from the sun.

These laws are the crowning glory, not only of Kepler's career, but of all astronomical discovery from the beginning up to his time, and they well deserve careful study and explanation, although more modern progress has shown that they are only approximately true.

EXERCISE 17.– Drive two pins into a smooth board an inch apart and fasten to them the ends of a string a foot long. Take up the slack of the string with the point of a lead pencil and, keeping the string drawn taut, move the pencil point over the board into every possible position. The curve thus traced will be an ellipse having the pins at the two points which are called its foci.

In the case of the planetary orbits one focus of the ellipse is vacant, and, in accordance with the first law, the center of the sun is at the other focus. In Fig. 17 the dot, inside the orbit of Mercury, which is marked *a*, shows the position of the vacant focus of the orbit of Mars, and the dot *b* is the vacant focus of Mercury's orbit. The orbits of Venus and the earth are so nearly circular that their vacant foci lie very close to the sun and are not marked in the figure. The line drawn from the sun to any point of the orbit (the string from pin to pencil point) is a *radius vector*. The point midway between the pins is the *center* of the ellipse, and the distance of either pin from the center measures the *eccentricity* of the ellipse.

Draw several ellipses with the same length of string, but with the pins at different distances apart, and note that the greater the eccentricity the flatter is the ellipse, but that all of them have the same length.

If both pins were driven into the same hole, what kind of an ellipse would you get?

The Second Law was worked out by Kepler as his answer to a problem suggested by the first law. In Fig. 17 it is apparent from a mere inspection of the orbit of Mercury that this planet travels much faster on one side of its orbit than on the other, the distance covered in ten days between the numbers 10 and 20 being more than fifty per cent greater than that between 50 and 60. The same difference is found, though usually in less degree, for every other planet, and Kepler's problem was to discover a means by which to mark upon the orbit the figures showing the positions of the planet at the end of equal intervals of time. His solution of this problem, contained in the second law, asserts that if we draw radii vectors from the sun to each of the marked points taken at equal time intervals around the orbit, then the area of the sector formed by two adjacent radii vectores and the arc included between them is equal to the area of each and every other such sector, the short radii vectores being spread apart so as to include a long arc between them while the long radii vectores have a short arc. In Kepler's form of stating the law the radius vector is supposed to travel with the planet and in each day to sweep over the same fractional part of the total area of the orbit. The spacing of the numbers in Fig. 17 was done by means of this law.

For the proper understanding of Kepler's Third Law we must note that the "mean distance" which appears in it is one half of the long diameter of the orbit and that the "periodic time" means the number of days or years required by the planet to make a complete circuit in its orbit. Representing the first of these by a and the second by T, we have, as the mathematical equivalent of the law,

$$a^3 \div T^2 = C$$

where the quotient, C, is a number which, as Kepler found, is the same for every planet of the solar system. If we take the mean distance of the earth from the sun as the unit of distance, and the year as the unit of time, we shall find by applying the equation to the earth's motion, $C = 1$. Applying this value to any other planet we shall find in the same units, $a = T^{2/3}$, by means of which we may determine the distance of any planet from the sun when its periodic time, T, has been learned from observation.

EXERCISE 18.– Uranus requires 84 years to make a revolution in its orbit. What is its mean distance from the sun? What are the mean distances of Mercury, Venus, and Mars? (See Chapter III for their periodic times.) Would it be possible for two planets at different distances from the sun to move around their orbits in the same time?

A circle is an ellipse in which the two foci have been brought together. Would Kepler's laws hold true for such an orbit?

34. ***Newton's laws of motion.***– Kepler studied and described the motion of the planets. Newton, three generations later (1727 A. D.), studied and described the mechanism which controls that motion. To Kepler and his age the heavens were supernatural, while to Newton and his successors they are a part of Nature, governed by the same laws which obtain upon the earth, and we turn to the ordinary things of everyday life as the foundation of celestial mechanics.

Every one who has ridden a bicycle knows that he can coast farther upon a level road if it is smooth than if it is rough; but however smooth and hard the road may be and however fast the wheel may have been started, it is sooner or later stopped by the resistance which the road and the air offer to its motion, and when once stopped or checked it can be started again only by applying fresh power. We have here a familiar illustration of what is called

The first law of motion.– "Every body continues in its state of rest or of uniform motion in a straight line except in so far as it may be compelled by force to change that state." A gust of wind, a stone, a careless movement of the rider may turn the bicycle to the right or the left, but unless some disturbing force is applied it will go straight ahead, and if all resistance to its motion could be removed it would go always at the speed given it by the last power applied, swerving neither to the one hand nor the other.

When a slow rider increases his speed we recognize at once that he has applied additional power to the wheel, and when this speed is slackened it equally shows that force has been applied against the motion. It is force alone which can produce a change in either velocity or direction of motion; but simple as this law now appears it required the genius of Galileo to discover it and of Newton to give it the form in which it is stated above.

35. ***The second law of motion***, which is also due to Galileo and Newton, is:

"Change of motion is proportional to force applied and takes place in the direction of the straight line in which the force acts." Suppose a man to fall from a balloon at some great elevation in the air; his own weight is the force which pulls him down, and that force operating at every instant is sufficient to give him at the end of the first second of his fall a downward velocity of 32 feet per second– i. e., it has changed his state from rest, to motion at this rate, and the motion is toward the earth because the force acts in that direction. During the next second the ceaseless operation of this force will have the same effect as in the first second and will add another 32 feet to his velocity, so that two seconds from the time he commenced to fall he will be moving at the rate of 64 feet per second, etc. The column of figures marked v in the table below shows what his velocity will be at the end of subsequent seconds. The changing velocity here shown is the change of motion to which the law refers, and the velocity is proportional to the time shown in the first column of the table, because the amount of force exerted in this case is proportional to the time during which it operated. The distance through which the man will fall in each second is shown in the column marked d, and is found by taking the average of his velocity at the beginning and end of this second, and the total distance through which he has fallen at the end of each second, marked s in the table, is found by taking the sum of all the preceding values of d. The velocity, 32 feet per second, which measures the change of motion in each second, also measures the *accelerating force* which produces this motion, and it is usually represented in formulæ by the letter g. Let the student show from the numbers in the table that the accelerating force, the time, t, during which it operates, and the space, s, fallen through, satisfy the relation

$$s = 1/2\, gt^2,$$

which is usually called the law of falling bodies. How does the table show that g is equal to 32?

Table

t	*v*	*d*	*s*
0	0	0	0
1	32	16	16
2	64	48	64
3	96	80	144
4	128	112	256
5	160	144	400
etc.	etc.	etc.	etc.

If the balloon were half a mile high how long would it take to fall to the ground? What would be the velocity just before reaching the ground?

Galileo Galilei (1564-1642)

Fig. 19 shows the path through the air of a ball which has been struck by a bat at the point *A*, and started off in the direction *A B* with a velocity of 200 feet per second. In accordance with the first law of motion, if it were acted upon by no other force than the impulse given by the bat, it should travel along the straight line *A B* at the uniform rate of 200 feet per second, and at the end of the fourth second it should be 800 feet from *A*, at the point marked 4, but during these four seconds its weight has caused it to fall 256 feet, and its actual position, 4', is 256 feet below the point 4. In this way we find its position at the end of each second, 1', 2', 3', 4', etc., and drawing a line through these points we shall find the actual path of the ball under the influence of the two forces to be the curved line *A C*. No matter how far the ball may go before striking the ground, it can not get back to the point *A*, and the curve *A C* therefore can not be a part of a circle, since that curve returns into itself. It is, in fact, a part of a *parabola*, which, as we shall see later, is a kind of orbit in which comets and some other heavenly bodies move. A skyrocket moves in the same kind of a path, and so does a stone, a bullet, or any other object hurled through the air.

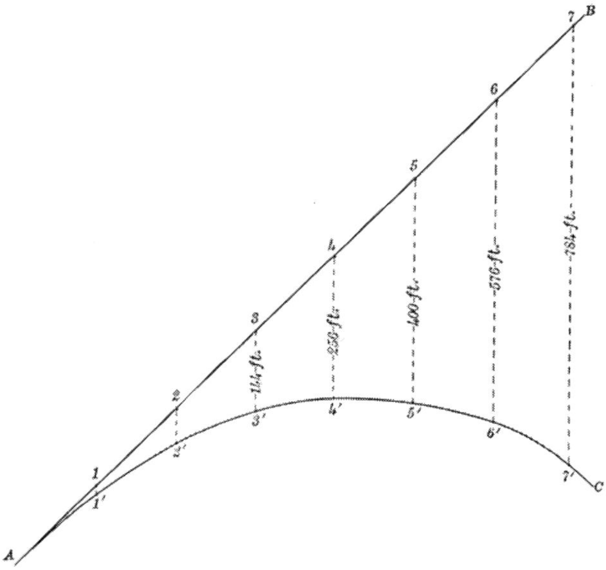

FIG. 19.—The path of a ball.

36. ***The third law of motion.***— "To every action there is always an equal and contrary reaction; or the mutual actions of any two bodies are always equal and oppositely directed." This is well illustrated in the case of a man climbing a rope hand over hand. The direct force or action which he exerts is a downward pull upon the rope, and it is the reaction of the rope to this pull which lifts him along it. We shall find in a later chapter a curious application of this law to the history of the earth and moon.

It is the great glory of Sir Isaac Newton that he first of all men recognized that these simple laws of motion hold true in the heavens as well as upon the earth; that the complicated motion of a planet, a comet, or a star is determined in accordance with these laws by the forces which act upon the bodies, and that these forces are essentially the same as that which we call weight. The formal statement of the principle last named is included in–

37. ***Newton's law of gravitation.***– "Every particle of matter in the universe attracts every other particle with a force whose direction is that of a line joining the two, and whose magnitude is directly as the product of their masses, and inversely as the square of their distance from each other." We know that we ourselves and the things about us are pulled toward the earth by a force (weight) which is called, in the Latin that Newton wrote, *gravitas*, and the word marks well the true significance of the law of gravitation. Newton did not discover a new force in the heavens, but he extended an old and familiar one from a limited terrestrial sphere of action to an unlimited and celestial one, and furnished a precise statement of the way in which the force operates. Whether a body be hot or cold, wet or dry, solid, liquid, or gaseous, is of no account in determining the force which it exerts, since this depends solely upon mass and distance.

The student should perhaps be warned against straining too far the language which it is customary to employ in this connection. The law of gravitation is certainly a far-reaching one, and it may operate in every remotest corner of the universe precisely as stated above, but additional information about those corners would be welcome to supplement our rather scanty stock of knowledge concerning what happens there. We may not controvert the words of a popular preacher who says, "When I lift my hand I move the stars in Ursa Major," but we should not wish to stand sponsor for them, even though they are justified by a rigorous interpretation of the Newtonian law.

The word *mass*, in the statement of the law of gravitation, means the quantity of matter contained in the body, and if we represent by the letters m' and m'' the respective quantities of matter contained in the two bodies whose distance from each other is r, we shall have, in accordance with the law of gravitation, the following mathematical expression for the force, F, which acts between them:

$$F = k\,(m'm'')/r^2.$$

This equation, which is the general mathematical expression for the law of gravitation, may be made to yield some curious results. Thus, if we select two bullets, each having a mass of 1 gram, and place them so that their centers are 1 centimeter apart, the above expression for the force exerted between them becomes

$$F = k\,\{(1 \times 1)/1^2\} = k,$$

from which it appears that the coefficient k is the force exerted between these bodies. This is called the gravitation constant, and it evidently furnishes a measure of the specific intensity with which one particle of matter attracts another. Elaborate experiments which have been made to determine the amount of this force show that it is surprisingly small, for in the case of the two bullets whose mass of 1 gram each is supposed to be concentrated into an indefinitely small space, gravity would have to operate between them continuously for more than forty minutes in order to pull them together, although they were separated by only 1 centimeter to start with, and nothing save their own inertia opposed their movements. It is only when one or both of the masses m', m'' are very great that the force of gravity becomes large, and the weight of bodies at the surface of the earth is considerable because of the great quantity of matter which goes to make up the earth. Many of the heavenly bodies are much more massive than the earth, as the mathematical astronomers have found by applying the law of gravitation to determine numerically their masses, or, in more popular language, to "weigh" them.

The student should observe that the two terms mass and weight are not synonymous; mass is defined above as the quantity of matter contained in a body, while weight is the force with which the earth attracts that body, and in accordance with the law of gravitation its weight depends upon its distance from the center of the earth, while its mass is quite independent of its position with respect to the earth.

By the third law of motion the earth is pulled toward a falling body just as strongly as the body is pulled toward the earth– i. e., by a force equal to the weight of the body. How much does the earth rise toward the body?

38. ***The motion of a planet.***– In Fig. 20 *S* represents the sun and *P* a planet or other celestial body, which for the moment is moving along the straight line *P 1*. In accordance with the first law of motion it would continue to move along this line with uniform velocity if no external force acted upon it; but such a force, the sun's attraction, is acting, and by virtue of this attraction the body is pulled aside from the line *P 1*.

Knowing the velocity and direction of the body's motion and the force with which the sun attracts it, the mathematician is able to apply Newton's laws of motion so as to determine the path of the body, and a few of the possible orbits are shown in the figure where the short cross stroke marks the point of each orbit which is nearest to the sun. This point is called the *perihelion*.

Without any formal application of mathematics we may readily see that the swifter the motion of the body at *P* the shorter will be the time during which it is subjected to the sun's attraction at close range, and therefore the force exerted by the sun, and the resulting change of motion, will be small, as in the orbits *P 1* and *P 2*.

On the other hand, *P 5* and *P 6* represent orbits in which the velocity at *P* was comparatively small, and the resulting change of motion greater than would be possible for a more swiftly moving body.

What would be the orbit if the velocity at *P* were reduced to nothing at all?

What would be the effect if the body starting at *P* moved directly away from *1*?

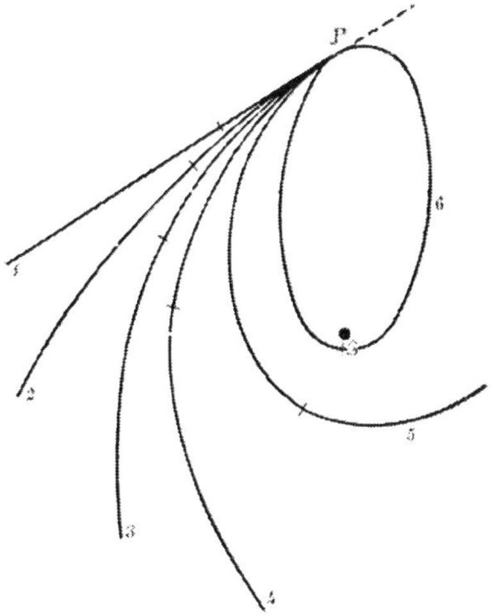

FIG. 20.— Different kinds of orbits.

The student should not fail to observe that the sun's attraction tends to pull the body at *P* forward along its path, and therefore increases its velocity, and that this influence continues until the planet reaches perihelion, at which point it attains its greatest velocity, and the force of the sun's attraction is wholly expended in changing the direction of its motion. After the planet has passed perihelion the sun begins to pull backward and to retard the motion in just the same measure that before perihelion passage it increased it, so that the two halves of the orbit on opposite sides of a line drawn from the perihelion through the sun are exactly alike. We may here note the explanation of Kepler's second law: when the planet is near the sun it moves faster, and the radius vector changes its direction more rapidly than when the planet is remote from the sun on account of the greater force with which it is attracted, and the exact relation between the rates at which the radius vector turns in different parts of the orbit, as given by the second law, depends upon the changes in this force.

When the velocity is not too great, the sun's backward pull, after a planet has passed perihelion, finally overcomes it and turns the planet toward the sun again, in such a way that it comes back to the point *P*, moving in the same direction and with the same speed as before— i. e., it has gone around the sun in an orbit like *P* 6 or *P* 4, an ellipse, along which it will continue to move ever after. But we must not fail to note that this return into the same orbit is a consequence of the last line in the statement of the law of gravitation (p. 54), and that, if the magnitude of this force were inversely as the cube of the distance or any other proportion than the square, the orbit would be something very different. If the velocity is too great for the sun's attraction to overcome, the orbit will be a hyperbola, like *P* 2, along which the body will move away never to return, while a velocity just at the limit of what the sun can control gives an orbit like *P* 3, a parabola, along which the body moves with *parabolic velocity*, which is ever diminishing as the body gets farther from the sun, but is always just sufficient to keep it from returning. If the earth's velocity could be increased 41 per cent, from 19 up to 27 miles per second, it would have parabolic velocity, and would quit the sun's company.

The summation of the whole matter is that the orbit in which a body moves around the sun, or past the sun, depends upon its velocity and if this velocity and the direction of the motion at any one point in the orbit are known the whole orbit is determined by them, and the position of the planet in its orbit for past as well as future times can be determined through the application of Newton's laws; and the same is true for any other heavenly body– moon, comet, meteor, etc. It is in this way that astronomers are able to predict, years in advance, in what particular part of the sky a given planet will appear at a given time.

It is sometimes a source of wonder that the planets move in ellipses instead of circles, but it is easily seen from Fig. 20 that the planet, *P*, could not by any possibility move in a circle, since the direction of its motion at *P* is not at right angles with the line joining it to the sun as it must be in a circular orbit, and even if it were perpendicular to the radius vector the planet must needs have exactly the right velocity given to it at this point, since either more or less speed would change the circle into an ellipse. In order to produce circular motion there must be a balancing of conditions as nice as is required to make a pin stand upon its point, and the really surprising thing is that the orbits of the planets should be so nearly circular as they are. If the orbit of the earth were drawn accurately to scale, the untrained eye would not detect the slightest deviation from a true circle, and even the orbit of Mercury (Fig. 17), which is much more eccentric than that of the earth, might almost pass for a circle.

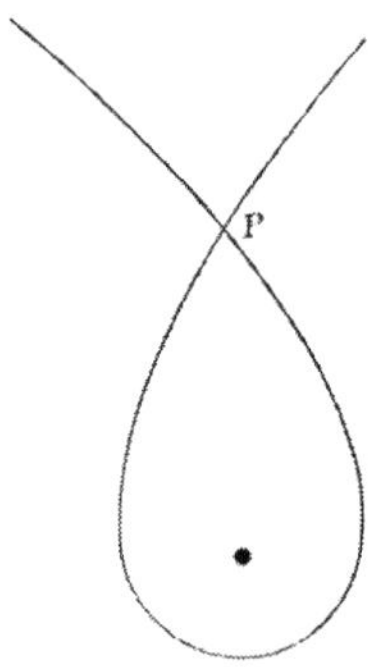

FIG. 21. An impossible orbit.

The orbit *P* 2, which lies between the parabola and the straight line, is called in geometry a hyperbola, and Newton succeeded in proving from the law of gravitation that a body might move under the sun's attraction in a hyperbola as well as in a parabola or ellipse; but it must move in some one of these curves; no other orbit is possible.[A] Thus it would not be possible for a body moving under the law of gravitation to describe about the sun any such orbit as is shown in Fig. 21. If the body passes a second time through any point of its orbit, such as *P* in the figure, then it must retrace, time after time, the whole path that it first traversed in getting from *P* around to *P* again– i. e., the orbit must be an ellipse.

Newton also proved that Kepler's three laws are mere corollaries from the law of gravitation, and that to be strictly correct the third law must be slightly altered so as to take into account the masses of the planets. These are, however, so small in comparison with that of the sun, that the correction is of comparatively little moment.

39. ***Perturbations.*** – In what precedes we have considered the motion of a planet under the influence of no other force than the sun's attraction, while in fact, as the law of gravitation asserts, every other body in the universe is in some measure attracting it and changing its motion. The resulting disturbances in the motion of the attracted body are called *perturbations*, but for the most part these are insignificant, because the bodies by whose disturbing attractions they are caused are either very small or very remote, and it is only when our moving planet, *P*, comes under the influence of some great disturbing power like Jupiter or one of the other planets that the perturbations caused by their influence need to be taken into account.

The problem of the motion of three bodies– sun, Jupiter, planet– which must then be dealt with is vastly more complicated than that which we have considered, and the ablest mathematicians and astronomers have not been able to furnish a complete solution for it, although they have worked upon the problem for two centuries, and have developed an immense amount of detailed information concerning it.

The Lick Observatory, Mount Hamilton, Cal.

In general each planet works ceaselessly upon the orbit of every other, changing its size and shape and position, backward and forward in accordance with the law of gravitation, and it is a question of serious moment how far this process may extend. If the diameter of the earth's orbit were very much increased or diminished by the perturbing action of the other planets, the amount of heat received from the sun would be correspondingly changed, and the earth, perhaps, be rendered unfit for the support of life. The tipping of the plane of the earth's orbit into a new position might also produce serious consequences; but the great French mathematician of a century ago, Laplace, succeeded in proving from the law of gravitation that although both of these changes are actually in progress they can not, at least for millions of years, go far enough to prove of serious consequence, and the same is true for all the other planets, unless here and there an asteroid may prove an exception to the rule.

The precession (Chapter V) is a striking illustration of a perturbation of slightly different character from the above, and another is found in connection with the plane of the moon's orbit. It will be remembered that the moon in its motion among the stars never goes far from the ecliptic, but in a complete circuit of the heavens crosses it twice, once in going from south to north and once in the opposite direction. The points at which it crosses the ecliptic are called the *nodes*, and under the perturbing influence of the sun these nodes move westward along the ecliptic about twenty degrees per year, an extraordinarily rapid perturbation, and one of great consequence in the theory of eclipses.

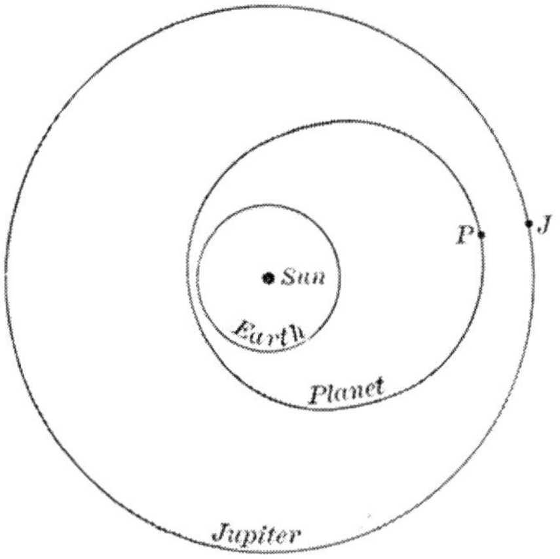

FIG. 22.—A planet subject to great perturbations by Jupiter.

40. ***Weighing the planets.***— Although these perturbations can not be considered dangerous, they are interesting since they furnish a method for weighing the planets which produce them. From the law of gravitation we learn that the ability of a planet to produce perturbations depends directly upon its mass, since the force F which it exerts contains this mass, m', as a factor. So, too, the divisor r^2 in the expression for the force shows that the

distance between the disturbing and disturbed bodies is a matter of great consequence, for the smaller the distance the greater the force. When, therefore, the mass of a planet such as Jupiter is to be determined from the perturbations it produces, it is customary to select some such opportunity as is presented in Fig. 22, where one of the small planets, called asteroids, is represented as moving in a very eccentric orbit, which at one point approaches close to the orbit of Jupiter, and at another place comes near to the orbit of the earth. For the most part Jupiter will not exert any very great disturbing influence upon a planet moving in such an orbit as this, since it is only at rare intervals that the asteroid and Jupiter approach so close to each other, as is shown in the figure. The time during which the asteroid is little affected by the attraction of Jupiter is used to study the motion given to it by the sun's attraction– that is, to determine carefully the undisturbed orbit in which it moves; but there comes a time at which the asteroid passes close to Jupiter, as shown in the figure, and the orbital motion which the sun imparts to it will then be greatly disturbed, and when the planet next comes round to the part of its orbit near the earth the effect of these disturbances upon its apparent position in the sky will be exaggerated by its close proximity to the earth. If now the astronomer observes the actual position of the asteroid in the sky, its right ascension and declination, and compares these with the position assigned to the planet by the law of gravitation when the attraction of Jupiter is ignored, the differences between the observed right ascensions and declinations and those computed upon the theory of undisturbed motion will measure the influence that Jupiter has had upon the asteroid, and the amount by which Jupiter has shifted it, compared with the amount by which the sun has moved it– that is, with the motion in its orbit– furnishes the mass of Jupiter expressed as a fractional part of the mass of the sun.

There has been determined in this manner the mass of every planet in the solar system which is large enough to produce any appreciable perturbation, and all these masses prove to be exceedingly small fractions of the mass of the sun, as may be seen from the following table, in which is given opposite the name of each planet the number by which the mass of the sun must be divided in order to get the mass of the planet:

Mercury	7,000,000 (?)
Venus	408,000
Earth	329,000
Mars	3,093,500
Jupiter	1,047.4

Saturn	3,502
Uranus	22,800
Neptune	19,700

It is to be especially noted that the mass given for each planet includes the mass of all the satellites which attend it, since their influence was felt in the perturbations from which the mass was derived. Thus the mass assigned to the earth is the combined mass of earth and moon.

41. ***Discovery of Neptune.***— The most famous example of perturbations is found in connection with the discovery, in the year 1846, of Neptune, the outermost planet of the solar system. For many years the motion of Uranus, his next neighbor, had proved a puzzle to astronomers. In accordance with Kepler's first law this planet should move in an ellipse having the sun at one of its foci, but no ellipse could be found which exactly fitted its observed path among the stars, although, to be sure, the misfit was not very pronounced. Astronomers surmised that the small deviations of Uranus from the best path which theory combined with observation could assign, were due to perturbations in its motion caused by an unknown planet more remote from the sun— a thing easy to conjecture but hard to prove, and harder still to find the unknown disturber. But almost simultaneously two young men, Adams in England and Le Verrier in France, attacked the problem quite independently of each other, and carried it to a successful solution, showing that if the irregularities in the motion of Uranus were indeed caused by an unknown planet, then that planet must, in September, 1846, be in the direction of the constellation Aquarius; and there it was found on September 23d by the astronomers of the Berlin Observatory whom Le Verrier had invited to search for it, and found within a degree of the exact point which the law of gravitation in his hands had assigned to it.

This working backward from the perturbations experienced by Uranus to the cause which produced them is justly regarded as one of the greatest scientific achievements of the human intellect, and it is worthy of note that we are approaching the time at which it may be repeated, for Neptune now behaves much as did Uranus three quarters of a century ago, and the most plausible explanation which can be offered for these anomalies in its path is that the bounds of the solar system must be again enlarged to include another disturbing planet.

42. ***The shape of a planet.***— There is an effect of gravitation not yet touched upon, which is of considerable interest and wide application in astronomy— viz., its influence in determining the shape of the heavenly bodies. The earth is a globe because every part of it is drawn toward the center by the attraction of the other parts, and if this attraction on its surface were

everywhere of equal force the material of the earth would be crushed by it into a truly spherical form, no matter what may have been the shape in which it was originally made. But such is not the real condition of the earth, for its diurnal rotation develops in every particle of its body a force which is sometimes called *centrifugal*, but which is really nothing more than the inertia of its particles, which tend at every moment to keep unchanged the direction of their motion and which thus resist the attraction that pulls them into a circular path marked out by the earth's rotation, just as a stone tied at the end of a string and swung swiftly in a circle pulls upon the string and opposes the constraint which keeps it moving in a circle. A few experiments with such a stone will show that the faster it goes the harder does it pull upon the string, and the same is true of each particle of the earth, the swiftly moving ones near the equator having a greater centrifugal force than the slow ones near the poles. At the equator the centrifugal force is directly opposed to the force of gravity, and in effect diminishes it, so that, comparatively, there is an excess of gravity at the poles which compresses the earth along its axis and causes it to bulge out at the equator until a balance is thus restored. As we have learned from the study of geography, in the case of the earth, this compression amounts to about 27 miles, but in the larger planets, Jupiter and Saturn, it is much greater, amounting to several thousand miles.

But rotation is not the only influence that tends to pull a planet out of shape. The attraction which the earth exerts upon the moon is stronger on the near side and weaker on the far side of our satellite than at its center, and this difference of attraction tends to warp the moon, as is illustrated in Fig. 23 where *1*, *2*, and *3* represent pieces of iron of equal mass placed in line on a table near a horseshoe magnet, *H*. Each piece of iron is attracted by the magnet and is held back by a weight to which it is fastened by means of a cord running over a pulley, *P*, at the edge of the table. These weights are all to be supposed equally heavy and each of them pulls upon its piece of iron with a force just sufficient to balance the attraction of the magnet for the middle piece, No. 2. It is clear that under this arrangement No. *2* will move neither to the right nor to the left, since the forces exerted upon it by the magnet and the weight just balance each other. Upon No. *1*, however, the magnet pulls harder than upon No. *2*, because it is nearer and its pull therefore more than balances the force exerted by the weight, so that No. *1* will be pulled away from No. *2* and will stretch the elastic cords, which are represented by the lines joining *1* and *2*, until their tension, together with the force exerted by the weight, just balances the attraction of the magnet. For No. *3*, the force exerted by the magnet is less than that of the weight, and it will also be pulled away from No. *2* until its elastic cords are stretched to the proper tension. The net result is that the three blocks which, without the magnet's influence, would be held close together by the elastic cords, are pulled apart by this outside force as far as the resistance of the cords will permit.

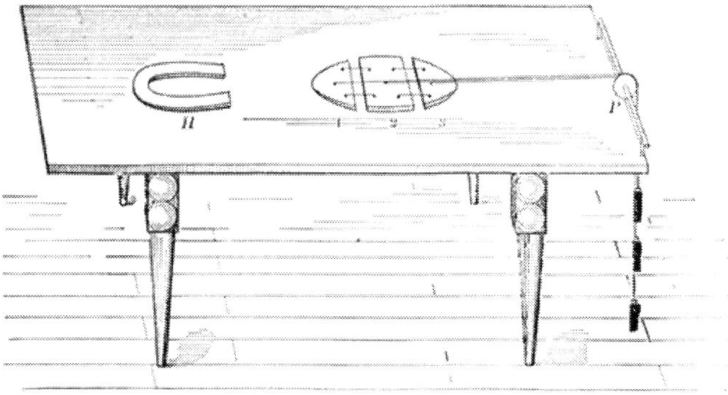

FIG. 23.— Tide-raising forces.

An entirely analogous set of forces produces a similar effect upon the shape of the moon. The elastic cords of Fig. 23 stand for the attraction of gravitation by which all the parts of the moon are bound together. The magnet represents the earth pulling with unequal force upon different parts of the moon. The weights are the inertia of the moon in its orbital motion which, as we have seen in a previous section, upon the whole just balances the earth's attraction and keeps the moon from falling into it. The effect of these forces is to stretch out the moon along a line pointing toward the earth, just as the blocks were stretched out along the line of the magnet, and to make this diameter of the moon slightly but permanently longer than the others.

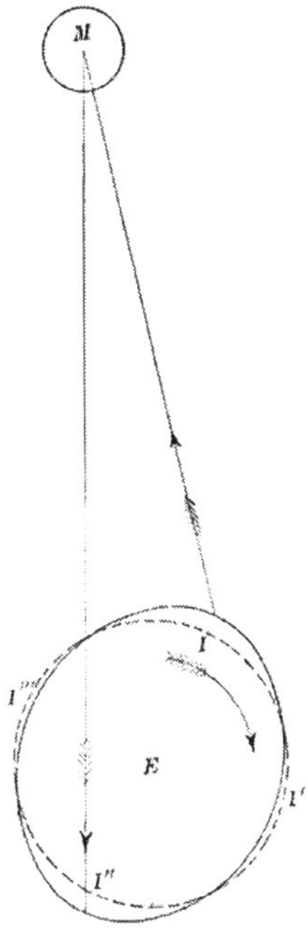

FIG. 24.— The tides.

The tides.— Similarly the moon and the sun attract opposite sides of the earth with different forces and feebly tend to pull it out of shape. But here a new element comes into play: the earth turns so rapidly upon its axis that its solid parts have no time in which to yield sensibly to the strains, which shift rapidly from one diameter to another as different parts of the earth are turned toward the moon, and it is chiefly the waters of the sea which respond to the distorting effect of the sun's and moon's attraction. These are heaped up on opposite sides of the earth so as to produce a slight elongation of its diameter, and Fig. 24 shows how by the earth's rotation this swelling of the waters is swept out from under the moon and is pulled back by the moon until it finally takes up some such position as that shown in the figure where the effect of the earth's rotation in carrying it one way is just balanced by the moon's attraction urging it back on line with the moon. This heaping up of the waters is called a *tide*. If *I* in the figure represents a little island in the sea the waters which surround it will of course accompany it in its diurnal rotation about the earth's axis, but whenever the island comes back to the position *I*, the waters will swell up as a part of the tidal wave and will encroach upon the land in what is called high tide or flood tide. So too when they reach *I''*, half a day later, they will again rise in flood tide, and midway between these points, at *I'*, the waters must subside, giving low or ebb tide.

The height of the tide raised by the moon in the open sea is only a very few feet, and the tide raised by the sun is even less, but along the coast of a continent, in bays and angles of the shore, it often happens that a broad but low tidal wave is forced into a narrow corner, and then the rise of the water may be many feet, especially when the solar tide and the lunar tide come in together, as they do twice in every month, at new and full moon. Why do they come together at these times instead of some other?

Small as are these tidal effects, it is worth noting that they may in certain cases be very much greater— e. g., if the moon were as massive as is the sun its tidal effect would be some millions of times greater than it now is and would suffice to grind the earth into fragments. Although the earth escapes this fate, some other bodies are not so fortunate, and we shall see in later chapters some evidence of their disintegration.

43. ***The scope of the law of gravitation.***— In all the domain of physical science there is no other law so famous as the Newtonian law of gravitation;

none other that has been so dwelt upon, studied, and elaborated by astronomers and mathematicians, and perhaps none that can be considered so indisputably proved. Over and over again mathematical analysis, based upon this law, has pointed out conclusions which, though hitherto unsuspected, have afterward been found true, as when Newton himself derived as a corollary from this law that the earth ought to be flattened at the poles– a thing not known at that time, and not proved by actual measurement until long afterward. It is, in fact, this capacity for predicting the unknown and for explaining in minutest detail the complicated phenomena of the heavens and the earth that constitutes the real proof of the law of gravitation, and it is therefore worth while to note that at the present time there are a very few points at which the law fails to furnish a satisfactory account of things observed. Chief among these is the case of the planet Mercury, the long diameter of whose orbit is slowly turning around in a way for which the law of gravitation as yet furnishes no explanation. Whether this is because the law itself is inaccurate or incomplete, or whether it only marks a case in which astronomers have not yet properly applied the law and traced out its consequences, we do not know; but whether it be the one or the other, this and other similar cases show that even here, in its most perfect chapter, astronomy still remains an incomplete science.

CHAPTER V.
THE EARTH AS A PLANET

44. ***The size of the earth.***— The student is presumed to have learned, in his study of geography, that the earth is a globe about 8,000 miles in diameter and, without dwelling upon the "proofs" which are commonly given for these statements, we proceed to consider the principles upon which the measurement of the earth's size and shape are based.

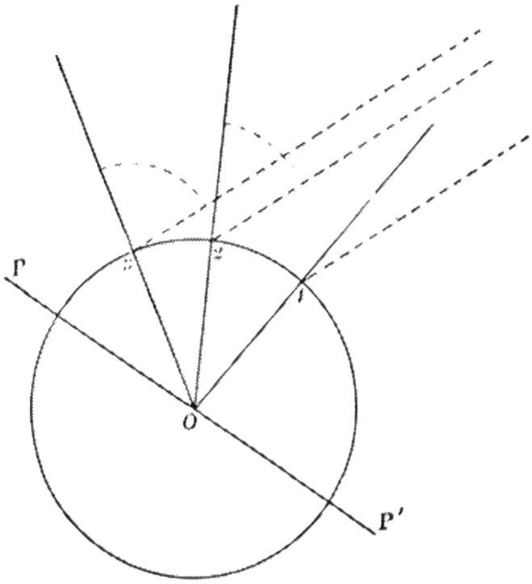

*FIG. 25.— **Measuring the size of the earth.***

In Fig. 25 the circle represents a meridian section of the earth; *P P'* is the axis about which it rotates, and the dotted lines represent a beam of light coming from a star in the plane of the meridian, and so distant that the dotted lines are all practically parallel to each other. The several radii drawn through the points *1*, *2*, *3*, represent the direction of the vertical at these points, and the angles which these radii produced, make with the rays of starlight are each equal to the angular distance of the star from the zenith of the place at the moment the star crosses the meridian. We have already seen, in Chapter II, how these angles may be measured, and it is apparent from the figure that the difference between any two of these angles— e. g., the angles at *1* and 2— is equal to the angle at the center, O, between the points *1* and 2. By measuring these angular distances of the star from the zenith, the astronomer finds the angles at the center of the earth between the stations *1*, 2, *3*, etc., at which his observations are made. If the meridian were a perfect circle the change of zenith distance of the star, as one traveled along a meridian from the equator to the pole, would be perfectly uniform— the same number of degrees for each hundred miles traveled— and observations made in many parts of the earth show that this is very nearly true, but that, on the whole, as we approach the pole it is necessary to travel a little greater distance than is required for a given change in the angle at the equator. The earth is, in fact, flattened at the poles to the amount of about 27 miles in the length of its diameter, and by this amount, as well as by smaller variations due to mountains and valleys, the shape of the earth differs from a perfect sphere. These astronomical measurements of the curvature of the earth's surface furnish by far the most satisfactory proof that it is very approximately a sphere, and furnish as its equatorial diameter 7,926 miles.

Neglecting the *compression*, as it is called, i. e., the 27 miles by which the equatorial diameter exceeds the polar, the size of the earth may easily be found by measuring the distance *1 - 2* along the surface and by combining with this the angle *1* O *2* obtained through measuring the meridian altitudes of any star as seen from *1* and *2*. Draw on paper an angle equal to the measured difference of altitude and find how far you must go from its vertex in order to have the distance between the sides, measured along an arc of a circle, equal to the measured distance between *1* and *2*. This distance from the vertex will be the earth's radius.

EXERCISE 19.– Measure the diameter of the earth by the method given above. In order that this may be done satisfactorily, the two stations at which observations are made must be separated by a considerable distance– i. e., 200 miles. They need not be on the same meridian, but if they are on different meridians in place of the actual distance between them, there must be used the projection of that distance upon the meridian– i. e., the north and south part of the distance.

By co-operation between schools in the Northern and Southern States, using a good map to obtain the required distances, the diameter of the earth may be measured with the plumb-line apparatus described in Chapter II and determined within a small percentage of its true value.

45. ***The mass of the earth.***– We have seen in Chapter IV the possibility of determining the masses of the planets as fractional parts of the sun's mass, but nothing was there shown, or could be shown, about measuring these masses after the common fashion in kilogrammes or tons. To do this we must first get the mass of the earth in tons or kilogrammes, and while the principles involved in this determination are simple enough, their actual application is delicate and difficult.

In Fig. 26 we suppose a long plumb line to be suspended above the surface of the earth and to be attracted toward the center of the earth, C, by a force whose intensity is (Chapter IV)

$F = k\, mE/R2$,

where E denotes the mass of the earth, which is to be determined by experiment, and R is the radius of the earth, 3,963 miles.

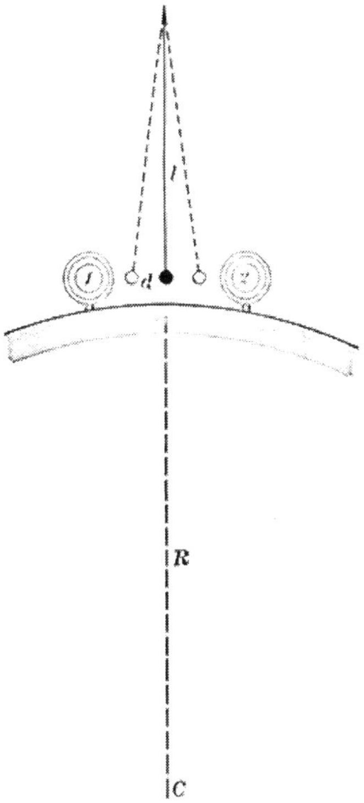

FIG. 26.— Illustrating the principles involved in weighing the earth.

If there is no disturbing influence present, the plumb line will point directly downward, but if a massive ball of lead or other heavy substance is placed at one side, *1*, it will attract the plumb line with a force equal to

$f = k\ mB/r^2$,

where r is the distance of its center from the plumb bob and B is its mass which we may suppose, for illustration, to be a ton. In consequence of this attraction the plumb line will be pulled a little to one side, as shown by the dotted line, and if we represent by l the length of the plumb line and by d the distance between the original and the disturbed positions of the plumb bob we may write the proportion

$F : f :: l : d$;

and introducing the values of F and f given above, and solving for E the proportion thus transformed, we find

$E = B \cdot l/d \cdot (R/r)^2$.

In this equation the mass of the ball, B, the length of the plumb line, l, the distance between the center of the ball and the center of the plumb bob, r, and the radius of the earth, R, can all be measured directly, and d, the amount by which the plumb bob is pulled to one side by the ball, is readily found by shifting the ball over to the other side, at 2, and measuring with a microscope how far the plumb bob moves. This distance will, of course, be equal to 2 d.

By methods involving these principles, but applied in a manner more complicated as well as more precise, the mass of the earth is found to be, in tons, $6,642 \times 10^{18}$– i. e., 6,642 followed by 18 ciphers, or in kilogrammes $60,258 \times 10^{20}$. The earth's atmosphere makes up about a millionth part of this mass.

If the length of the plumb line were 100 feet, the weight of the ball a ton, and the distance between the two positions of the ball, *1* and *2*, six feet, how many inches, d, would the plumb bob be pulled out of place?

Find from the mass of the earth and the data of § 40 the mass of the sun in tons. Find also the mass of Mars. The computation can be very greatly abridged by the use of logarithms.

46. ***Precession.***— That the earth is isolated in space and has no support upon which to rest, is sufficiently shown by the fact that the stars are visible upon every side of it, and no support can be seen stretching out toward them. We must then consider the earth to be a globe traveling freely about the sun in a circuit which it completes once every year, and rotating once in every twenty-four hours about an axis which remains at all seasons directed very nearly toward the star Polaris. The student should be able to show from his own observations of the sun that, with reference to the stars, the direction of the sun from the earth changes about a degree a day. Does this prove that the earth revolves about the sun?

But it is only in appearance that the pole maintains its fixed position among the stars. If photographs are taken year after year, after the manner of Exercise 7, it will be found that slowly the pole is moving (nearly) toward Polaris, and making this star describe a smaller and smaller circle in its diurnal path, while stars on the other side of the pole (in right ascension 12h.) become more distant from it and describe larger circles in their diurnal motion; but the process takes place so slowly that the space of a lifetime is required for the motion of the pole to equal the angular diameter of the full moon.

Spin a top and note how its rapid whirl about its axis corresponds to the earth's diurnal rotation. When the axis about which the top spins is truly vertical the top "sleeps"; but if the axis is tipped ever so little away from the vertical it begins to wobble, so that if we imagine the axis prolonged out to the sky and provided with a pencil point as a marker, this would trace a circle around the zenith, along which the pole of the top would move, and a little observation will show that the more the top is tipped from the vertical the larger does this circle become and the more rapidly does the wobbling take place. Were it not for the spinning of the top about its axis, it would promptly fall over when tipped from the vertical position, but the spin combines with the force which pulls the top over and produces the wobbling motion. Spin the top in opposite directions, with the hands of a watch and contrary to the hands of a watch, and note the effect which is produced upon the wobbling.

The earth presents many points of resemblance to the top. Its diurnal rotation is the spin about the axis. This axis is tipped 23.5° away from the perpendicular to its orbit (obliquity of the ecliptic) just as the axis of the top is tipped away from the vertical line. In consequence of its rapid spin, the body of the earth bulges out at the equator (27 miles), and the sun and moon, by virtue of their attraction (see Chapter IV), lay hold of this protuberance and pull it down toward the plane of the earth's orbit, so that if it were not for the spin this force would straighten the axis up and set it perpendicular to the orbit plane. But here, as in the case of the top, the spin and the tipping force combine to produce a wobble which is called precession, and whose effect we recognize in the shifting position of the pole among the stars. The motion of precession is very much slower than the wobbling of the top, since the tipping force for the earth is relatively very small, and a period of nearly 26,000 years is required for a complete circuit of the pole about its center of motion. Friction ultimately stops both the spin and the wobble of the top, but this influence seems wholly absent in the case of the earth, and both rotation and precession go on unchanged from century to century, save for certain minor forces which for a time change the direction or rate of the precessional motion, first in one way and then in another, without in the long run producing any results of consequence.

The center of motion, about which the pole travels in a small circle having an angular radius of 23.5°, is at that point of the heavens toward which a perpendicular to the plane of the earth's orbit points, and may be found on the star map in right ascension 18h. 0m. and declination 66.5°.

EXERCISE 20.– Find this point on the map, and draw as well as you can the path of the pole about it. The motion of the pole along its path is toward the constellation Cepheus. Mark the position of the pole along this path at intervals of 1,000 years, and refer to these positions in dealing with some of the following questions:

Does the wobbling of the top occur in the same direction as the motion of precession? Do the tipping forces applied to the earth and top act in the same direction? What will be the polar star 12,000 years hence? The Great Pyramid of Egypt is thought to have been used as an observatory when Alpha Draconis was the bright star nearest the pole. How long ago was that?

The motion of the pole of course carries the equator and the equinoxes with it, and thus slowly changes the right ascensions and declinations of all the stars. On this account it is frequently called the precession of the equinoxes, and this motion of the equinox, slow though it is, is a matter of some consequence in connection with chronology and the length of the year.

Will the precession ever bring back the right ascensions and declinations to be again what they now are?

In what direction is the pole moving with respect to the Big Dipper? Will its motion ever bring it exactly to Polaris? How far away from Polaris will the precession carry the pole? What other bright stars will be brought near the pole by the precession?

47. ***The warming of the earth.***– Winter and summer alike the day is on the average warmer than the night, and it is easy to see that this surplus of heat comes from the sun by day and is lost by night through radiation into the void which surrounds the earth; just as the heat contained in a mass of molten iron is radiated away and the iron cooled when it is taken out from the furnace and placed amid colder surroundings. The earth's loss of heat by radiation goes on ceaselessly day and night, and were it not for the influx of solar heat this radiation would steadily diminish the temperature toward what is called the "absolute zero"– i. e., a state in which all heat has been taken

away and beyond which there can be no greater degree of cold. This must not be confounded with the zero temperatures shown by our thermometers, since it lies nearly 500° below the zero of the Fahrenheit scale (-273° Centigrade), a temperature which by comparison makes the coldest winter weather seem warm, although the ordinary thermometer may register many degrees below its zero. The heat radiated by the sun into the surrounding space on every side of it is another example of the same cooling process, a hot body giving up its heat to the colder space about it, and it is the minute fraction of this heat poured out by the sun, and in small part intercepted by the earth, which warms the latter and produces what we call weather, climate, the seasons, etc.

Observe the fluctuations, the ebb and flow, which are inherent in this process. From sunset to sunrise there is nothing to compensate the steady outflow of heat, and air and ground grow steadily colder, but with the sunrise there comes an influx of solar heat, feeble at first because it strikes the earth's surface very obliquely, but becoming more and more efficient as the sun rises higher in the sky. But as the air and the ground grow warm during the morning hours they part more and more readily and rapidly with their store of heat, just as a steam pipe or a cup of coffee radiates heat more rapidly when very hot. The warmest hour of the day is reached when these opposing tendencies of income and expenditure of heat are just balanced; and barring such disturbing factors as wind and clouds, the gain in temperature usually extends to the time— an hour or two beyond noon— at which the diminishing altitude of the sun renders his rays less efficient, when radiation gains the upper hand and the temperature becomes for a short time stationary, and then commences to fall steadily until the next sunrise.

We have here an example of what is called a periodic change— i. e., one which, within a definite and uniform period (24 hours), oscillates from a minimum up to a maximum temperature and then back again to a minimum, repeating substantially the same variation day after day. But it must be understood that minor causes not taken into account above, such as winds, water, etc., produce other fluctuations from day to day which sometimes obscure or even obliterate the diurnal variation of temperature caused by the sun.

Expose the back of your hand to the sun, holding the hand in such a position that the sunlight strikes perpendicularly upon it; then turn the hand so that the light falls quite obliquely upon it and note how much more vigorous is the warming effect of the sun in the first position than in the second. It is chiefly this difference of angle that makes the sun's warmth more effective when he is high up in the sky than when he is near the horizon, and more effective in summer than in winter.

We have seen in Chapter III that the sun's motion among the stars takes place along a path which carries it alternately north and south of the equator to a distance of 23.5°, and the stars show by their earlier risings and later settings, as we pass from the equator toward the north pole of the heavens, that as the sun moves northward from the equator, each day in the northern hemisphere will become a little longer, each night a little shorter, and every day the sun will rise higher toward the zenith until this process culminates toward the end of June, when the sun begins to move southward, bringing shorter days and smaller altitudes until the Christmas season, when again it is reversed and the sun moves northward. We have here another periodic variation, which runs its complete course in a period of a year, and it is easy to see that this variation must have a marked effect on the warming of the earth, the long days and great altitudes of summer producing the greater warmth of that season, while the shorter days and lower altitudes of December, by diminishing the daily supply of solar heat, bring on the winter's cold. The succession of the seasons, winter following summer and summer winter, is caused by the varying altitude of the sun, and this in turn is due to the obliquity of the ecliptic, or, what is the same thing, the amount by which the axis of the earth is tipped from being perpendicular to the plane of its orbit, and the seasons are simply a periodic change in the warming of the earth, quite comparable with the diurnal change but of longer period.

It is evident that the period within which the succession of winter and summer is completed, the year, as we commonly call it, must equal the time required by the sun to go from the vernal equinox around to the vernal equinox again, since this furnishes a complete cycle of the sun's motions north and south from the equator. On account of the westward motion of the equinox (precession) this is not quite the same as the time required for a complete revolution of the earth in its orbit, but is a little shorter (20m. 23s.), since the equinox moves back to meet the sun.

48. ***Relation of the sun to climate.***— It is clear that both the northern and southern hemispheres of the earth must have substantially the same kind of seasons, since the motion of the sun north and south affects both alike; but when the sun is north of the equator and warming our hemisphere most effectively, his light falls more obliquely upon the other hemisphere, the days there are short and winter reigns at the time we are enjoying summer, while six months later the conditions are reversed.

In those parts of the earth near the equator— the torrid zone— there is no such marked change from cold to warm as we experience, because, as the sun never gets more than 23.5° away from the celestial equator, on every day of the year he mounts high in the tropic skies, always coming within 23.5° of the zenith, and usually closer than this, so that there is no such periodic change in the heat supply as is experienced in higher latitudes, and within the tropics the temperature is therefore both higher and more uniform than in our latitude.

In the frigid zones, on the contrary, the sun never rises high in the sky; at the poles his greatest altitude is only 23.5°, and during the winter season he does not rise at all, so that the temperature is here low the whole year round, and during the winter season, when for weeks or months at a time the supply of solar light is entirely cut off, the temperature falls to a degree unknown in more favored climes.

If the obliquity of the ecliptic were made 10° greater, what would be the effect upon the seasons in the temperate zones? What if it were made 10° less?

Does the precession of the equinoxes have any effect upon the seasons or upon the climate of different parts of the earth?

If the axis of the earth pointed toward Arcturus instead of Polaris, would the seasons be any different from what they are now?

49. ***The atmosphere.***— Although we live upon its surface, we are not outside the earth, but at the bottom of a sea of air which forms the earth's outermost layer and extends above our heads to a height of many miles. The study of most of the phenomena of the atmosphere belongs to that branch of physics called meteorology, but there are a few matters which fairly come within our consideration of the earth as a planet. We can not see the stars save as we look through this atmosphere, and the light which comes through it is bent and oftentimes distorted so as to present serious obstacles to any accurate telescopic study of the heavenly bodies. Frequently this disturbance is visible to the naked eye, and the stars are said to twinkle— i. e., to quiver and change color many times per second, solely in consequence of a disturbed condition of the air and not from anything which goes on in the star. This effect is more marked low down in the sky than near the zenith, and it is worth noting that the planets show very little of it because the light they send to the earth comes from a disk of sensible area, while a star, being much smaller and farther from the earth, has its disk reduced practically to a mere point whose light is more easily affected by local disturbances in the atmosphere than is the broader beam which comes from the planets' disk.

50. ***Refraction.***— At all times, whether the stars twinkle or not, their light is bent in its passage through the atmosphere, so that the stars appear to stand higher up in the sky than their true positions. This effect, which the astronomer calls refraction, must be allowed for in observations of the more precise class, although save at low altitudes its amount is a very small fraction of a degree, but near the horizon it is much exaggerated in amount and becomes easily visible to the naked eye by distorting the disks of the sun and moon from circles into ovals with their long diameters horizontal. The refraction lifts both upper and lower edge of the sun, but lifts the lower edge more than the upper, thus shortening the vertical diameter. See Fig. 27, which shows not only this effect, but also the reflection of the sun from the curved surface of the sea, still further flattening the image. If the surface of the water

were flat, the reflected image would have the same shape as the sun's disk, and its altered appearance is sometimes cited as a proof that the earth's surface is curved.

The total amount of the refraction at the horizon is a little more than half a degree, and since the diameters of the sun and moon subtend an angle of about half a degree, we have the remarkable result that in reality the whole disk of either sun or moon is below the horizon at the instant that the lower edge appears to touch the horizon and sunset or moonset begins. The same effect exists at sunrise, and as a consequence the duration of sunshine or of moonshine is on the average about six minutes longer each day than it would be if there were no atmosphere and no refraction. A partial offset to this benefit is found in the fact that the atmosphere absorbs the light of the heavenly bodies, so that stars appear much less bright when near the horizon than when they are higher up in the sky, and by reason of this absorption the setting sun can be looked at with the naked eye without the discomfort which its dazzling luster causes at noon.

*FIG. 27.— **Flattening of the sun's disk by refraction and by reflection from the surface of the sea.***

51. ***The twilight.***— Another effect of the atmosphere, even more marked than the preceding, is the twilight. As at sunrise the mountain top catches the rays of the coming sun before they reach the lowland, and at sunset it keeps them after they have faded from the regions below, so the particles of dust and vapor, which always float in the atmosphere, catch the sunlight and reflect it to the surface of the earth while the sun is still below the horizon, giving at the beginning and end of day that vague and diffuse light which we call twilight.

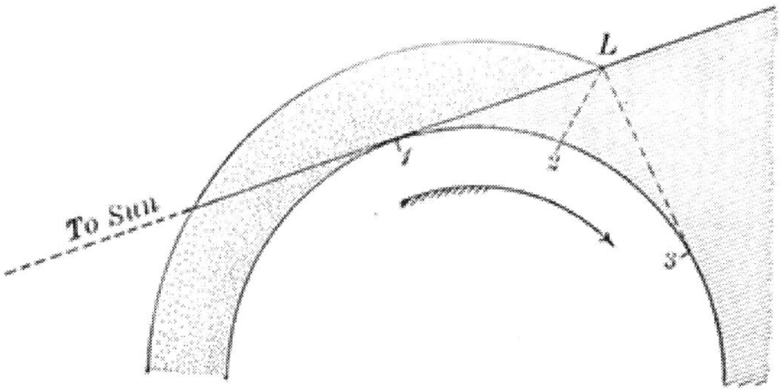

FIG. 28.— Twilight phenomena.

Fig. 28 shows a part of the earth surrounded by such a dust-laden atmosphere, which is illuminated on the left by the rays of the sun, but which, on the right of the figure, lies in the shadow cast by the earth. To an observer placed at *1* the sun is just setting, and all the atmosphere above him is illumined with its rays, which furnish a bright twilight. When, by the earth's rotation, this observer has been carried to 2, all the region to the east of his zenith lies in the shadow, while to the west there is a part of the atmosphere from which there still comes a twilight, but now comparatively faint, because the lower part of the atmosphere about our observer lies in the shadow, and it is mainly its upper regions from which the light comes, and here the dust and moisture are much less abundant than in the lower strata. Still later, when the observer has been carried by the earth's rotation to the point *3*, every vestige of twilight will have vanished from his sky, because all of the illuminated part of the atmosphere is now below his horizon, which is represented by the line *3 L*. In the figure the sun is represented to be 78° below this horizon line at the end of twilight, but this is a gross exaggeration, made for the sake of clearness in the drawing— in fact, twilight is usually said to end when the sun is 18° below the horizon.

Let the student redraw Fig. 28 on a large scale, so that the points *1* and *3* shall be only 18° apart, as seen from the earth's center. He will find that the point *L* is brought down much closer to the surface of the earth, and measuring the length of the line *2 L*, he should find for the "height of the atmosphere" about one-eightieth part of the radius of the earth— i. e., a little less than 50 miles. This, however, is not the true height of the atmosphere. The air extends far beyond this, but the particles of dust and vapor which are capable of sending sunlight down to the earth seem all to lie below this limit.

The student should not fail to watch the eastern sky after sunset, and see the shadow of the earth rise up and fill it while the twilight arch retreats steadily toward the west.

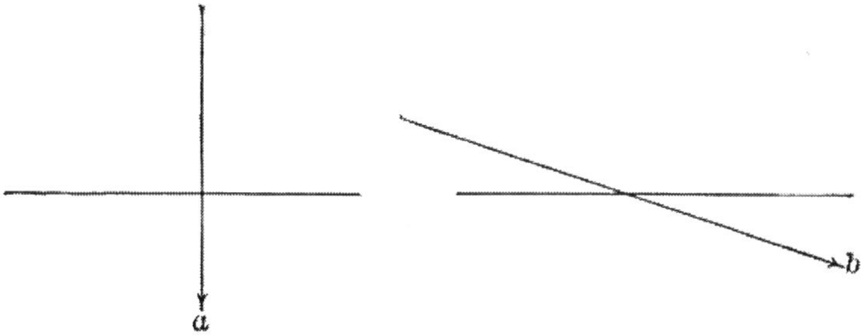

FIG. 29.— The cause of long and short twilights.

Duration of twilight.— Since twilight ends when the sun is 18° below the horizon, any circumstance which makes the sun go down rapidly will shorten the duration of twilight, and anything which retards the downward motion of the sun will correspondingly prolong it. Chief among influences of this kind is the angle which the sun's course makes with the horizon. If it goes straight down, as at *a*, Fig. 29, a much shorter time will suffice to carry it to a depression of 18° than is needed in the case shown at *b* in the same figure, where the motion is very oblique to the horizon. If we consider different latitudes and different seasons of the year, we shall find every possible variety of circumstance from *a* to *b*, and corresponding to these, the duration of twilight varies from an all-night duration in the summers of Scotland and more northern lands to an hour or less in the mountains of Peru. For the sake of graphical effect, the shortness of tropical twilight is somewhat exaggerated by Coleridge in the lines,

"The sun's rim dips; the stars rush out:At one stride comes the dark." *The Ancient Mariner.*

In the United States the longest twilights come at the end of June, and last for a little more than two hours, while the shortest ones are in March and September, amounting to a little more than an hour and a half; but at all times the last half hour of twilight is hardly to be distinguished from night, so small is the quantity of reflecting matter in the upper regions of the atmosphere. For practical convenience it is customary to assume in the courts of law that twilight ends an hour after sunset.

How long does twilight last at the north pole?

The Aurora.— One other phenomenon of the atmosphere may be mentioned, only to point out that it is not of an astronomical character. The Aurora, or northern lights, is as purely an affair of the earth as is a thunderstorm, and its explanation belongs to the subject of terrestrial magnetism.

CHAPTER VI.
THE MEASUREMENT OF TIME

52. ***Solar time.***— To measure any quantity we need a unit in terms of which it must be expressed. Angles are measured in degrees, and the degree is the unit for angular measurement. For most scientific purposes the centimeter is adopted as the unit with which to measure distances, and similarly a day is the fundamental unit for the measurement of time. Hours, minutes, and seconds are aliquot parts of this unit convenient for use in dealing with shorter periods than a day, and the week, month, and year which we use in our calendars are multiples of the day.

Strictly speaking, a day is not the time required by the earth to make one revolution upon its axis, but it is best defined as the amount of time required for a particular part of the sky to make the complete circuit from the meridian of a particular place through west and east back to the meridian again. The day begins at the moment when this specified part of the sky is on the meridian, and "the time" at any moment is the hour angle of this particular part of the sky— i. e., the number of hours, minutes, etc., that have elapsed since it was on the meridian.

The student has already become familiar with the kind of day which is based upon the motion of the vernal equinox, and which furnishes sidereal time, and he has seen that sidereal time, while very convenient in dealing with the motions of the stars, is decidedly inconvenient for the ordinary affairs of life since in the reckoning of the hours it takes no account of daylight and darkness. One can not tell off-hand whether 10 hours, sidereal time, falls in the day or in the night. We must in some way obtain a day and a system of time reckoning based upon the apparent diurnal motion of the sun, and we may, if we choose, take the sun itself as the point in the heavens whose transit over the meridian shall mark the beginning and the end of the day. In this system "the time" is the number of hours, minutes, etc., which have elapsed since the sun was on the meridian, and this is the kind of time which is shown by a sun dial, and which was in general use, years ago, before clocks and watches became common. Since the sun moves among the stars about a degree per day, it is easily seen that the rotating earth will have to turn farther in order to carry any particular meridian from the sun around to the sun again, than to carry it from a star around to the same star, or from the vernal equinox around to the vernal equinox again; just as the minute hand of a clock turns farther in going from the hour hand round to the hour hand again than it turns in going from XII to XII. These solar days and hours and minutes are therefore a little longer than the corresponding sidereal ones, and this furnishes the explanation why the stars come to the meridian a little earlier, by solar time, every night than on the night before, and why sidereal time gains steadily upon solar time, this gain amounting to approximately 3m.

56.5s. per day, or exactly one day per year, since the sun makes the complete circuit of the constellations once in a year.

With the general introduction of clocks and watches into use about a century ago this kind of solar time went out of common use, since no well-regulated clock could keep the time correctly. The earth in its orbital motion around the sun goes faster in some parts of its orbit than in others, and in consequence the sun appears to move more rapidly among the stars in winter than in summer; moreover, on account of the convergence of hour circles as we go away from the equator, the same amount of motion along the ecliptic produces more effect in winter and summer when the sun is north or south, than it does in the spring and autumn when the sun is near the equator, and as a combined result of these causes and other minor ones true solar time, as it is called, is itself not uniform, but falls behind the uniform lapse of sidereal time at a variable rate, sometimes quicker, sometimes slower. A true solar day, from noon to noon, is 51 seconds shorter in September than in December.

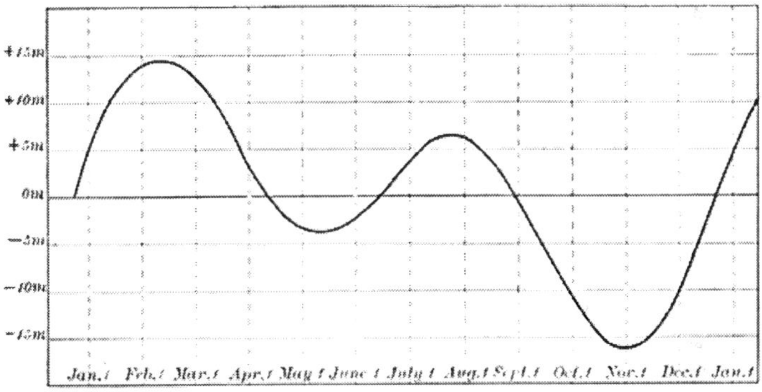

FIG. 30.— The equation of time.

53. ***Mean solar time.***— To remedy these inconveniences there has been invented and brought into common use what is called *mean solar time*, which is perfectly uniform in its lapse and which, by comparison with sidereal time, loses exactly one day per year. "The time" in this system never differs much from true solar time, and the difference between the two for any particular day may be found in any good almanac, or may be read from the curve in Fig. 30, in which the part of the curve above the line marked *0m* shows how many minutes mean solar time is faster than true solar time. The correct name for this difference between the two kinds of solar time is the *equation of time*, but in the almanacs it is frequently marked "sun fast" or "sun slow." In sidereal time and true solar time the distinction between A. M. hours (*ante meridiem* = before the sun reaches the meridian) and P. M. hours (*post meridiem* = after the sun has passed the meridian) is not observed, "the time"

being counted from 0 hours to 24 hours, commencing when the sun or vernal equinox is on the meridian. Occasionally the attempt is made to introduce into common use this mode of reckoning the hours, beginning the day (date) at midnight and counting the hours consecutively up to 24, when the next date is reached and a new start made. Such a system would simplify railway time tables and similar publications; but the American public is slow to adopt it, although the system has come into practical use in Canada and Spain.

54. ***To find (approximately) the sidereal time at any moment.*** — RULE I. When the mean solar time is known. Let *W* represent the time shown by an ordinary watch, and represent by *S* the corresponding sidereal time and by *D* the number of days that have elapsed from March 23d to the date in question. Then

$$S = W + 69/70 \times D \times 4.$$

The last term is expressed in minutes, and should be reduced to hours and minutes. Thus at 4 P. M. on July 4th—

D	=	103 days.
69/70 × D × 4	=	406m.
	=	6h. 46m.
W	=	4h. 0m.
S	=	10h. 46m.

The daily gain of sidereal upon mean solar time is 69/70 of 4 minutes, and March 23d is the date on which sidereal and mean solar time are together, taking the average of one year with another, but it varies a little from year to year on account of the extra day introduced in leap years.

RULE II. When the stars in the northern sky can be seen. Find β Cassiopeiæ, and imagine a line drawn from it to Polaris, and another line from Polaris to the zenith. The sidereal time is equal to the angle between these lines, provided that that angle must be measured from the zenith toward the west. Turn the angle from degrees into hours by dividing by 15.

55. ***The earth's rotation.*** — We are familiar with the fact that a watch may run faster at one time than at another, and it is worth while to inquire if the same is not true of our chief timepiece— the earth. It is assumed in the sections upon the measurement of time that the earth turns about its axis with absolute uniformity, so that mean solar time never gains or loses even the smallest fraction of a second. Whether this be absolutely true or not, no one has ever succeeded in finding convincing proof of a variation large enough to be measured, although it has recently been shown that the axis about which it rotates is not perfectly fixed within the body of the earth. The solid body

of the earth wriggles about this axis like a fish upon a hook, so that the position of the north pole upon the earth's surface changes within a year to the extent of 40 or 50 feet (15 meters) without ever getting more than this distance away from its average position. This is probably caused by the periodical shifting of masses of air and water from one part of the earth to another as the seasons change, and it seems probable that these changes will produce some small effect upon the rotation of the earth. But in spite of these, for any such moderate interval of time as a year or a century, so far as present knowledge goes, we may regard the earth's rotation as uniform and undisturbed. For longer intervals— e. g., 1,000,000 or 10,000,000 years— the question is a very different one, and we shall have to meet it again in another connection.

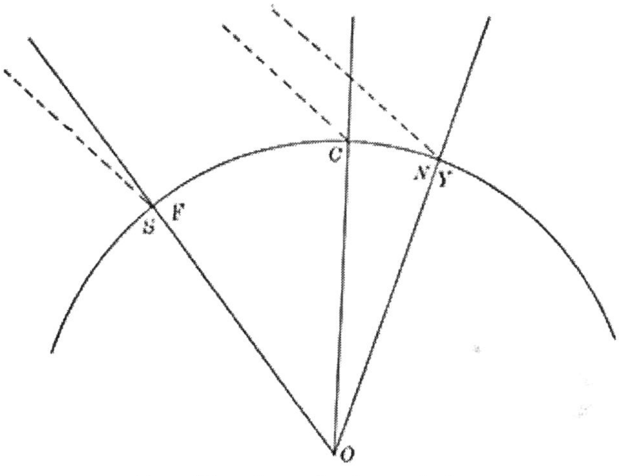

FIG. 31.— Longitude and time

56. ***Longitude and time.***— In what precedes there has been constant reference to the meridian. The day begins when the sun is on the meridian. Solar time is the angular distance of the sun past the meridian. Sidereal time was determined by observing transits of stars over a meridian line actually laid out upon the ground, etc. But every place upon the earth has its own meridian from which "the time" may be reckoned, and in Fig. 31, where the rays of sunlight are represented as falling upon a part of the earth's equator through which the meridians of New York, Chicago, and San Francisco pass, it is evident that these rays make different angles with the meridians, and that the sun is farther from the meridian of New York than from that of San Francisco by an amount just equal to the angle at O between these meridians. This angle is called by geographers the difference of longitude between the

two places, and the student should note that the word longitude is here used in a different sense from that on page 36. From Fig. 31 we obtain the

Theorem.— The difference between "the times" at any two meridians is equal to their difference of longitude, and the time at the eastern meridian is greater than at the western meridian. Astronomers usually express differences of longitude in hours instead of degrees. 1h. = 15°.

The name given to any kind of time should distinguish all the elements which enter into it— e. g., New York sidereal time means the hour angle of the vernal equinox measured from the meridian of New York, Chicago true solar time is the hour angle of the sun reckoned from the meridian of Chicago, etc.

FIG. 32.— Standard time.

57. ***Standard time.***— The requirements of railroad traffic have led to the use throughout the United States and Canada of four "standard times," each of which is a mean solar time some integral number of hours slower than the time of the meridian passing through the Royal Observatory at Greenwich, England.

Eastern	Time is	hours	slower	than	that	of Greenwich.
Central	"	"	"	"	"	"
Mountain	"	"	"	"	"	"

Pacific " " " " " "

In Fig. 32 the broken lines indicate roughly the parts of the United States and Canada in which these several kinds of time are used, and illustrate how irregular are the boundaries of these parts.

Standard time is sent daily into all of the more important telegraph offices of the United States, and serves to regulate watches and clocks, to the almost complete exclusion of local time.

58. ***To determine the longitude.***— With an ordinary watch observe the time of the sun's transit over your local meridian, and correct the observed time for the equation of time by means of the curve in Fig. 30. The difference between the corrected time and 12 o'clock will be the correction of your watch referred to local mean solar time. Compare your watch with the time signals in the nearest telegraph office and find its correction referred to standard time. The difference between the two corrections is the difference between your longitude and that of the standard meridian.

N. B.— Don't tamper with the watch by trying to "set it right." No harm will be done if it is wrong, provided you take due account of the correction as indicated above.

If the correction of the watch changed between your observation and the comparison in the telegraph office, what effect would it have upon the longitude determination? How can you avoid this effect?

59. ***Chronology.***— The Century Dictionary defines chronology as "the science of time"— that is, "the method of measuring or computing time by regular divisions or periods according to the revolutions of the sun or moon."

We have already seen that for the measurement of short intervals of time the day and its subdivisions— hours, minutes, seconds— furnish a very complete and convenient system. But for longer periods, extending to hundreds and thousands of days, a larger unit of time is required, and for the most part these longer units have in all ages and among all peoples been based upon astronomical considerations. But to this there is one marked exception. The week is a simple multiple of the day, as the dime is a multiple of the cent, and while it may have had its origin in the changing phases of the moon this is at best doubtful, since it does not follow these with any considerable accuracy. If the still longer units of time— the month and the year— had equally been made to consist of an integral number of days much confusion and misunderstanding might have been avoided, and the annals of ancient times would have presented fewer pitfalls to the historian than is now the case. The month is plainly connected with the motion of the moon among the stars. The year is, of course, based upon the motion of the sun through the heavens and the change of seasons which is thus produced; although, as commonly employed, it is not quite the same as the time required by the earth to make one complete revolution in its orbit. This time of one revolution is called a sidereal year, while, as we have already seen in Chapter V, the year which measures the course of the seasons is shorter than this on account of the precession of the equinoxes. It is called a tropical year with reference to the circuit which the sun makes from one tropic to the other and back again.

We can readily understand why primitive peoples should adopt as units of time these natural periods, but in so doing they incurred much the same kind of difficulty that we should experience in trying to use both English and American money in the ordinary transactions of life. How many dollars make a pound sterling? How shall we make change with English shillings and American dimes, etc.? How much is one unit worth in terms of the other?

One of the Greek poets[B] has left us a quaint account of the confusion which existed in his time with regard to the place of months and moons in the calendar:

"The moon by us to you her greeting sends,But bids us say that she's an ill-used moonAnd takes it much amiss that you will stillShuffle her days and turn them topsy-turvy,So that when gods, who know their feast days well,By your false count are sent home supperless,They scold and storm at her for your neglect."

60. ***Day, month, and year.***— If the day, the month, and the year are to be used concurrently, it is necessary to determine how many days are contained in the month and year, and when this has been done by the astronomer the numbers are found to be very awkward and inconvenient for daily use; and much of the history of chronology consists in an account of the various devices by which ingenious men have sought to use integral numbers to replace the cumbrous decimal fractions which follow.

According to Professor Harkness, for the epoch 1900 A. D.—

One	tropical	year	=	365.242197 mean solar days.
"	"	"	=	365d. 5h. 48m. 45.8s.

One	lunation	=	29.530588 mean solar days.
"	"	=	29d. 12h. 44m. 2.8s.

The word *lunation* means the average interval from one new moon to the next one— i. e., the time required by the moon to go from conjunction with the sun round to conjunction again.

A very ancient device was to call a year equal to 365 days, and to have months alternately of 29 and 30 days in length, but this was unsatisfactory in more than one way. At the end of four years this artificial calendar would be about one day ahead of the true one, at the end of forty years ten days in error, and within a single lifetime the seasons would have appreciably changed their position in the year, April weather being due in March, according to the calendar. So, too, the year under this arrangement did not consist of any integral number of months, 12 months of the average length of 29.5 days being 354 days, and 13 months 383.5 days, thus making any particular month change its position from the beginning to the middle and the end of the year within a comparatively short time. Some peoples gave up the astronomical year as an independent unit and adopted a conventional year of 12 lunar months, 354 days, which is now in use in certain Mohammedan countries, where it is known as the wandering year, with reference to the changing positions of the seasons in such a year. Others held to the astronomical year and adopted a system of conventional months, such that twelve of them would just make up a year, as is done to this day in our own calendar, whose months of arbitrary length we are compelled to remember by some such jingle as the following:

"Thirty days hath September,April, June, and November;All the rest have thirty-oneSave February,Which alone hath twenty-eight,Till leap year gives it twenty-nine."

61. ***The calendar.*** – The foundations of our calendar may fairly be ascribed to Julius Cæsar, who, under the advice of the Egyptian astronomer Sosigines, adopted the old Egyptian device of a leap year, whereby every fourth year was to consist of 366 days, while ordinary years were only 365 days long. He also placed the beginning of the year at the first of January, instead of in March, where it had formerly been, and gave his own name, Julius, to the month which we now call July. August was afterward named in honor of his successor, Augustus. The names of the earlier months of the year are drawn from Roman mythology; those of the later months, September, October, etc., meaning seventh month, eighth month, represent the places of these months in the year, before Cæsar's reformation, and also their places in some of the subsequent calendars, for the widest diversity of practice existed during mediæval times with regard to the day on which the new year should begin, Christmas, Easter, March 25th, and others having been employed at different times and places.

The system of leap years introduced by Cæsar makes the average length of a year 365.25 days, which differs by about eleven minutes from the true length of the tropical year, a difference so small that for ordinary purposes no better approximation to the true length of the year need be desired. But *any* deviation from the true length, however small, must in the course of time shift the seasons, the vernal and autumnal equinox, to another part of the year, and the ecclesiastical authorities of mediæval Europe found here ground for objection to Cæsar's calendar, since the great Church festival of Easter has its date determined with reference to the vernal equinox, and with the lapse of centuries Easter became more and more displaced in the calendar, until Pope Gregory XIII, late in the sixteenth century, decreed another reformation, whereby ten days were dropped from the calendar, the day after March 11th being called March 21st, to bring back the vernal equinox to the date on which it fell in A. D. 325, the time of the Council of Nicæa, which Gregory adopted as the fundamental epoch of his calendar.

The calendar having thus been brought back into agreement with that of old time, Gregory purposed to keep it in such agreement for the future by modifying Cæsar's leap-year rule so that it should run: Every year whose number is divisible by 4 shall be a leap year except those years whose numbers are divisible by 100 but not divisible by 400. These latter years– e. g., 1900– are counted as common years. The calendar thus altered is called Gregorian to distinguish it from the older, Julian calendar, and it found speedy acceptance in those civilized countries whose Church adhered to Rome; but the Protestant powers were slow to adopt it, and it was introduced into England and her American colonies by act of Parliament in the year 1752, nearly two centuries after Gregory's time. In Russia the Julian calendar has remained in common use to our own day, but in commercial affairs it is there customary to write the date according to both calendars– e. g., July 4/16, and at the present time strenuous exertions are making in that country for the adoption of the Gregorian calendar to the complete exclusion of the Julian one.

The Julian and Gregorian calendars are frequently represented by the abbreviations O. S. and N. S., old style, new style, and as the older historical dates are usually expressed in O. S., it is sometimes convenient to transform a date from the one calendar to the other. This is readily done by the formula

$$G = J + (N - 2) - N/4,$$

where G and J are the respective dates, N is the number of the century, and the remainder is to be neglected in the division by 4. For September 3, 1752, O. S., we have

J	=	Sept. 3
N - 2	=	+ 15
- N/4	=	- 4
G	=	Sept. 14

and September 14 is the date fixed by act of Parliament to correspond to September 3, 1752, O. S. Columbus discovered America on October 12, 1492, O. S. What is the corresponding date in the Gregorian calendar?

62. ***The day of the week.***– A problem similar to the above but more complicated consists in finding the day of the week on which any given date of the Gregorian calendar falls– e. g., October 21, 1492.

The formula for this case is

$$7q + r = Y + D + (Y - 1)/4 - (Y - 1)/100 + (Y - 1)/400$$

where Y denotes the given year, D the number of the day (date) in that year, and q and r are respectively the quotient and the remainder obtained by dividing the second member of the equation by 7. If $r = 1$ the date falls on Sunday, etc., and if $r = 0$ the day is Saturday. For the example suggested above we have

Jan.	31
Feb.	29
Mch.	31
April	30
May	31
June	30
July	31
Aug.	31
Sept.	30
Oct.	21
D =	295

Y		=	1492	
+ D		=	+ 295	
+ (Y - 1) ÷	4	=	+ 372	
- (Y - 1) ÷	100	=	- 14	
+ (Y - 1) ÷	400	=	+ 3	
			7) 2148	
	q	=	306	
	r	=	6	= Friday.

Find from some history the day of the week on which Columbus first saw America, and compare this with the above.

On what day of the week did last Christmas fall? On what day of the week were you born? In the formula for the day of the week why does q have the coefficient 7? What principles in the calendar give rise to the divisors 4, 100, 400?

For much curious and interesting information about methods of reckoning the lapse of time the student may consult the articles Calendar and Chronology in any good encyclopædia.

The Yerkes Observatory, Williams Bay, Wis.

CHAPTER VII.
ECLIPSES

63. ***The nature of eclipses.***— Every planet has a shadow which travels with the planet along its orbit, always pointing directly away from the sun, and cutting off from a certain region of space the sunlight which otherwise would fill it. For the most part these shadows are invisible, but occasionally one of them falls upon a planet or some other body which shines by reflected sunlight, and, cutting off its supply of light, produces the striking phenomenon which we call an eclipse. The satellites of Jupiter, Saturn, and Mars are eclipsed whenever they plunge into the shadows cast by their respective planets, and Jupiter himself is partially eclipsed when one of his own satellites passes between him and the sun, and casts upon his broad surface a shadow too small to cover more than a fraction of it.

But the eclipses of most interest to us are those of the sun and moon, called respectively solar and lunar eclipses. In Fig. 33 the full moon, *M'*, is shown immersed in the shadow cast by the earth, and therefore eclipsed, and in the same figure the new moon, *M*, is shown as casting its shadow upon the earth and producing an eclipse of the sun. From a mere inspection of the figure we may learn that an eclipse of the sun can occur only at new moon— i. e., when the moon is on line between the earth and sun— and an eclipse of the moon can occur only at full moon. Why? Also, the eclipsed moon, *M'*, will present substantially the same appearance from every part of the earth where it is at all visible— the same from North America as from South America— but the eclipsed sun will present very different aspects from different parts of the earth. Thus, at *L*, within the moon's shadow, the sunlight will be entirely cut off, producing what is called a total eclipse. At points of the earth's surface near *J* and *K* there will be no interference whatever with the sunlight, and no eclipse, since the moon is quite off the line joining these regions to any part of the sun. At places between *J* and *L* or *K* and *L* the moon will cut off a part of the sun's light, but not all of it, and will produce what is called a partial eclipse, which, as seen from the northern parts of the earth, will be an eclipse of the lower (southern) part of the sun, and as seen from the southern hemisphere will be an eclipse of the northern part of the sun.

FIG. 33.— Different kinds of eclipse.

The moon revolves around the earth in a plane, which, in the figure, we suppose to be perpendicular to the surface of the paper, and to pass through the sun along the line *M' M* produced. But it frequently happens that this plane is turned to one side of the sun, along some such line as *P Q*, and in this case the full moon would cut through the edge of the earth's shadow without being at any time wholly immersed in it, giving a partial eclipse of the moon, as is shown in the figure.

In what parts of the earth would this eclipse be visible? What kinds of solar eclipse would be produced by the new moon at *Q*? In what parts of the earth would they be visible?

64. ***The shadow cone.***— The shape and position of the earth's shadow are indicated in Fig. 33 by the lines drawn tangent to the circles which represent the sun and earth, since it is only between these lines that the earth interferes with the free radiation of sunlight, and since both sun and earth are spheres, and the earth is much the smaller of the two, it is evident that the earth's shadow must be, in geometrical language, a cone whose base is at the earth, and whose vertex lies far to the right of the figure— in other words, the earth's shadow, although very long, tapers off finally to a point and ends. So, too, the shadow of the moon is a cone, having its base at the moon and its vertex turned away from the sun, and, as shown in the figure, just about long enough to reach the earth.

It is easily shown, by the theorem of similar triangles in connection with the known size of the earth and sun, that the distance from the center of the earth to the vertex of its shadow is always equal to the distance of the earth from the sun divided by 108, and, similarly, that the length of the moon's shadow is equal to the distance of the moon from the sun divided by 400, the moon's shadow being the smaller and shorter of the two, because the moon is smaller than the earth. The radius of the moon's orbit is just about 1/400th part of the radius of the earth's orbit— i. e., the distance of the moon from the earth is 1/400th part of the distance of the earth from the sun, and it is this "chance" agreement between the length of the moon's shadow and the distance of the moon from the earth which makes the tip of the moon's shadow fall very near the earth at the time of solar eclipses. Indeed, the elliptical shape of the moon's orbit produces considerable variations in the distance of the moon from the earth, and in consequence of these variations the vertex of the shadow sometimes falls short of reaching the earth, and sometimes even projects considerably beyond its farther side. When the moon's distance is too great for the shadow to bridge the space between earth and moon there can be no total eclipse of the sun, for there is no shadow which can fall upon the earth, even though the moon does come directly between earth and sun. But there is then produced a peculiar kind of partial eclipse called *annular*, or ring-shaped, because the moon, although eclipsing the central parts of the sun, is not large enough to cover the whole of it, but leaves the sun's edge visible as a ring of light, which completely surrounds the moon. Although, strictly speaking, this is only a partial eclipse, it is customary to put total and annular eclipses together in one class, which is called central eclipses, since in these eclipses the line of centers of sun and moon strikes the earth, while in ordinary partial eclipses it passes to one side of the earth without striking it. In this latter case we have to consider another cone called the *penumbra*— i. e., partial shadow— which is shown in Fig. 33 by the broken lines tangent to the sun and moon, and crossing at the point *V*, which is the vertex of this cone. This penumbral cone includes within its surface all that region of space within which the moon cuts off any of the sunlight, and of course it includes the shadow cone which produces total eclipses. Wherever the penumbra falls there will be a solar eclipse of some kind, and the nearer the place is to the axis of the penumbra, the more nearly total will be the eclipse. Since the moon stands about midway between the earth and the vertex of the penumbra, the diameter of the penumbra where it strikes the earth will be about twice as great as the diameter of the moon, and the student should be able to show from this that the region of the earth's surface within which a partial solar eclipse is visible extends in a straight line about 2,100 miles on either side of the region where the eclipse is total. Measured along the curved surface of the earth, this distance is frequently much greater.

Is it true that if at any time the axis of the shadow cone comes within 2,100 miles of the earth's surface a partial eclipse will be visible in those parts of the earth nearest the axis of the shadow?

65. ***Different characteristics of lunar and solar eclipses.*** — One marked difference between lunar and solar eclipses which has been already suggested, may be learned from Fig. 33. The full moon, *M'*, will be seen eclipsed from every part of the earth where it is visible at all at the time of the eclipse— that is, from the whole night side of the earth; while the eclipsed sun will be seen eclipsed only from those parts of the day side of the earth upon which the moon's shadow or penumbra falls. Since the point of the shadow at best but little more than reaches to the earth, the amount of space upon the earth which it can cover at any one moment is very small, seldom more than 100 to 200 miles in length, and it is only within the space thus actually covered by the shadow that the sun is at any given moment totally eclipsed, but within this region the sun disappears, absolutely, behind the solid body of the moon, leaving to view only such outlying parts and appendages as are too large for the moon to cover. At a lunar eclipse, on the other hand, the earth coming between sun and moon cuts off the light from the latter, but, curiously enough, does not cut it off so completely that the moon disappears altogether from sight even in mid-eclipse. The explanation of this continued visibility is furnished by the broken lines extending, in Fig. 33, from the earth through the moon. These represent sunlight, which, entering the earth's atmosphere

near the edge of the earth (edge as seen from sun and moon), passes through it and emerges in a changed direction, refracted, into the shadow cone and feebly illumines the moon's surface with a ruddy light like that often shown in our red sunsets. Eclipse and sunset alike show that when the sun's light shines through dense layers of air it is the red rays which come through most freely, and the attentive observer may often see at a clear sunset something which corresponds exactly to the bending of the sunlight into the shadow cone; just before the sun reaches the horizon its disk is distorted from a circle into an oval whose horizontal diameter is longer than the vertical one (see § 50).

QUERY.– At a total lunar eclipse what would be the effect upon the appearance of the moon if the atmosphere around the edge of the earth were heavily laden with clouds?

66. ***The track of the shadow.***– We may regard the moon's shadow cone as a huge pencil attached to the moon, moving with it along its orbit in the direction of the arrowhead (Fig. 34), and as it moves drawing a black line across the face of the earth at the time of total eclipse. This black line is the path of the shadow and marks out those regions within which the eclipse will be total at some stage of its progress. If the point of the shadow just reaches the earth its trace will have no sensible width, while, if the moon is nearer, the point of the cone will be broken off, and, like a blunt pencil, it will draw a broad streak across the earth, and this under the most favorable circumstances may have a breadth of a little more than 160 miles and a length of 10,000 or 12,000 miles. The student should be able to show from the known distance of the moon (240,000 miles) and the known interval between consecutive new moons (29.5 days) that on the average the moon's shadow sweeps past the earth at the rate of 2,100 miles per hour, and that in a general way this motion is from west to east, since that is the direction of the moon's motion in its orbit. The actual velocity with which the moon's shadow moves past a given station may, however, be considerably greater or less than this, since on the one hand when the shadow falls very obliquely, as when the eclipse occurs near sunrise or sunset, the shifting of the shadow will be very much greater than the actual motion of the moon which produces it, and on the other hand the earth in revolving upon its axis carries the spectator and the ground upon which he stands along the same direction in which the shadow is moving. At the equator, with the sun and moon overhead, this motion of the earth subtracts about 1,000 miles per hour from the velocity with which the shadow passes by. It is chiefly on this account, the diminished velocity with which the shadow passes by, that total solar eclipses last longer in the tropics than in higher latitudes, but even under the most favorable circumstances the duration of totality does not reach eight minutes at any one place, although it may take the shadow several hours to sweep the entire length of its path across the earth.

According to Whitmell the greatest possible duration of a total solar eclipse is 7m. 40s., and it can attain this limit only when the eclipse occurs near the beginning of July and is visible at a place 5° north of the equator.

The duration of a lunar eclipse depends mainly upon the position of the moon with respect to the earth's shadow. If it strikes the shadow centrally, as at *M'*, Fig. 33, a total eclipse may last for about two hours, with an additional hour at the beginning and end, during which the moon is entering and leaving the earth's shadow. If the moon meets the shadow at one side of the axis, as at *P*, the total phase of the eclipse may fail altogether, and between these extremes the duration of totality may be anything from two hours downward.

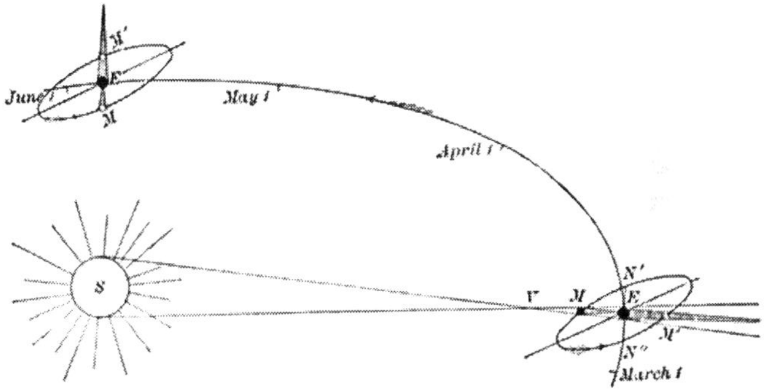

FIG. 34.— Relation of the lunar nodes to eclipses.

67. ***Relation of the lunar nodes to eclipses.***— To show why the moon sometimes encounters the earth's shadow centrally and more frequently at full moon passes by without touching it at all, we resort to Fig. 34, which represents a part of the orbit of the earth about the sun, with dates showing the time in each year at which the earth passes the part of its orbit thus marked. The orbit of the moon about the earth, *M M'*, is also shown, with the new moon, *M*, casting its shadow toward the earth and the full moon, *M'*, apparently immersed in the earth's shadow. But here appearances are deceptive, and the student who has made the observations set forth in Chapter III has learned for himself a fact of which careful account must now be taken. The apparent paths of the moon and sun among the stars are great circles which lie near each other, but are not exactly the same; and since these great circles are only the intersections of the sky with the planes of the earth's orbit and the moon's orbit, we see that these planes are slightly inclined to each other and must therefore intersect along some line passing through the center of the earth. This line, *N' N''*, is shown in the figure, and if we suppose the surface of the paper to represent the plane of the earth's orbit, we shall have to suppose the moon's orbit to be tipped around this line, so that the left side of the orbit lies above and the right side below the surface

of the paper. But since the earth's shadow lies in the plane of its orbit— i. e., in the surface of the paper— the full moon of March, *M'*, must have passed below the shadow, and the new moon, *M*, must have cast its shadow above the earth, so that neither a lunar nor a solar eclipse could occur in that month. But toward the end of May the earth and moon have reached a position where the line *N' N''* points almost directly toward the sun, in line with the shadow cones which hide it. Note that the line *N' N''* remains very nearly parallel to its original position, while the earth is moving along its orbit. The full moon will now be very near this line and therefore very close to the plane of the earth's orbit, if not actually in it, and must pass through the shadow of the earth and be eclipsed. So also the new moon will cast its shadow in the plane of the ecliptic, and this shadow, falling upon the earth, produced the total solar eclipse of May 28, 1900.

N' N'' is called the line of nodes of the moon's orbit (§ 39), and the two positions of the earth in its orbit, diametrically opposite each other, at which *N' N''* points exactly toward the sun, we shall call the *nodes* of the lunar orbit. Strictly speaking, the nodes are those points of the sky against which the moon's center is projected at the moment when in its orbital motion it cuts through the plane of the earth's orbit. Bearing in mind these definitions, we may condense much of what precedes into the proposition: Eclipses of either sun or moon can occur only when the earth is at or near one of the nodes of the moon's orbit. Corresponding to these positions of the earth there are in each year two seasons, about six months apart, at which times, and at these only, eclipses can occur. Thus in the year 1900 the earth passed these two points on June 2d and November 24th respectively, and the following list of eclipses which occurred in that year shows that all of them were within a few days of one or the other of these dates:

Eclipses of the Year 1900

Total solar eclipse	May 28th.
Partial lunar eclipse	June 12th.
Annular (solar) eclipse	November 21st.

68. ***Eclipse limits.***— If the earth is exactly at the node at the time of new moon, the moon's shadow will fall centrally upon it and will produce an eclipse visible within the torrid zone, since this is that part of the earth's surface nearest the plane of its orbit. If the earth is near but not at the node, the new moon will stand a little north or south of the plane of the earth's orbit, and its shadow will strike the earth farther north or south than before, producing an eclipse in the temperate or frigid zones; or the shadow may even

pass entirely above or below the earth, producing no eclipse whatever, or at most a partial eclipse visible near the north or south pole. Just how many days' motion the earth may be away from the node and still permit an eclipse is shown in the following brief table of eclipse limits, as they are called:

Solar Eclipse Limits

If at any new moon the earth is

Less than 10 days away	from	a	node,	a central eclipse is certain.
Between 10 and 16 days	"	"	"	some kind of eclipse is certain.
Between 16 and 19 days	"	"	"	a partial eclipse is possible.
More than 19 days	"	"	"	no eclipse is possible.

Lunar Eclipse Limits

If at any full moon the earth is

Less than 4 days away	from	a	node,	a total eclipse is certain.
Between 4 and 10 days	"	"	"	some kind of eclipse is certain.
Between 10 and 14 days	"	"	"	a partial eclipse is possible.
More than 14 days	"	"	"	no eclipse is possible.

From this table of eclipse limits we may draw some interesting conclusions about the frequency with which eclipses occur.

69. ***Number of eclipses in a year.*** – Whenever the earth passes a node of the moon's orbit a new moon must occur at some time during the 2 × 16 days that the earth remains inside the limits where some kind of eclipse is certain, and there must therefore be an eclipse of the sun every time the earth passes a node of the moon's orbit. But, since there are two nodes past which the earth moves at least once in each year, there must be at least two solar

eclipses every year. Can there be more than two? On the average, will central or partial eclipses be the more numerous?

A similar line of reasoning will not hold true for eclipses of the moon, since it is quite possible that no full moon should occur during the 20 days required by the earth to move past the node from the western to the eastern limit. This omission of a full moon while the earth is within the eclipse limits sometimes happens at both nodes in the same year, and then we have a year with no eclipse of the moon. The student may note in the list of eclipses for 1900 that the partial lunar eclipse of June 12th occurred 10 days after the earth passed the node, and was therefore within the doubtful zone where eclipses may occur and may fail, and corresponding to this position the eclipse was a very small one, only a thousandth part of the moon's diameter dipping into the shadow of the earth. By so much the year 1900 escaped being an illustration of a year in which no lunar eclipse occurred.

A partial eclipse of the moon will usually occur about a fortnight before or after a total eclipse of the sun, since the full moon will then be within the eclipse limit at the opposite node. A partial eclipse of the sun will always occur about a fortnight before or after a total eclipse of the moon.

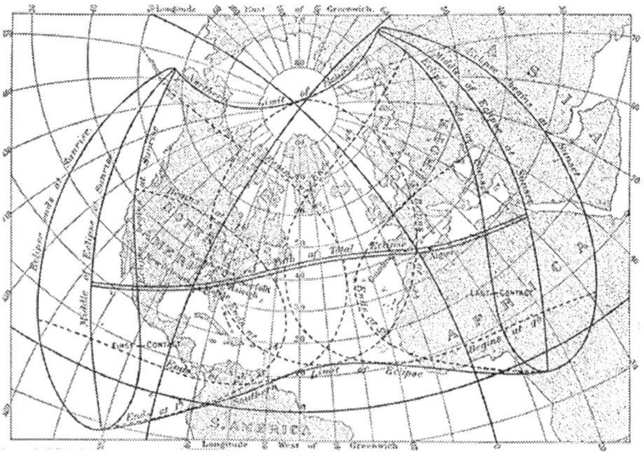

FIG. 35.— The eclipse of May 28, 1900.

70. ***Eclipse maps.***— It is the custom of astronomers to prepare, in advance of the more important eclipses, maps showing the trace of the moon's shadow across the earth, and indicating the times of beginning and ending of the eclipses, as is shown in Fig. 35. While the actual construction of such a map requires much technical knowledge, the principles involved are simple enough: the straight line passed through the center of sun and moon is the axis of the shadow cone, and the map contains little more than a graphical representation of when and where this cone meets the surface of the earth. Thus in the map, the "Path of Total Eclipse" is the trace of the shadow cone across the face of the earth, and the width of this path shows

that the earth encountered the shadow considerably inside the vertex of the cone. The general direction of the path is from west to east, and the slight sinuousities which it presents are for the most part due to unavoidable distortion of the map caused by the attempt to represent the curved surface of the earth upon the flat surface of the paper. On either side of the Path of Total Eclipse is the region within which the eclipse was only partial, and the broken lines marked Begins at 3h., Ends at 3h., show the intersection of the penumbral cone with the surface of the earth at 3 P. M., Greenwich time. These two lines inclose every part of the earth's surface from which at that time any eclipse whatever could be seen, and at this moment the partial eclipse was just beginning at every point on the eastern edge of the penumbra and just ending at every point on the western edge, while at the center of the penumbra, on the Path of Total Eclipse, lay the shadow of the moon, an oval patch whose greatest diameter was but little more than 60 miles in length, and within which lay every part of the earth where the eclipse was total at that moment.

The position of the penumbra at other hours is also shown on the map, although with more distortion, because it then meets the surface of the earth more obliquely, and from these lines it is easy to obtain the time of beginning and end of the eclipse at any desired place, and to estimate by the distance of the place from the Path of Total Eclipse how much of the sun's face was obscured.

Let the student make these "predictions" for Washington, Chicago, London, and Algiers.

The points in the map marked First Contact, Last Contact, show the places at which the penumbral cone first touched the earth and finally left it. According to computations made as a basis for the construction of the map the Greenwich time of First Contact was 0h. 12.5m. and of Last Contact 5h. 35.6m., and the difference between these two times gives the total duration of the eclipse upon the earth– i. e., 5 hours 23.1 minutes.

71. ***Future eclipses.***– An eclipse map of a different kind is shown in Fig. 36, which represents the shadow paths of all the central eclipses of the sun, visible during the period 1900-1918 A. D., in those parts of the earth north of the south temperate zone. Each continuous black line shows the path of the shadow in a total eclipse, from its beginning, at sunrise, at the western end of the line to its end, sunset, at the eastern end, the little circle near the middle of the line showing the place at which the eclipse was total at noon. The broken lines represent similar data for the annular eclipses. This map is one of a series prepared by the Austrian astronomer, Oppolzer, showing the path of every such eclipse from the year 1200 B. C. to 2160 A. D., a period of more than three thousand years.

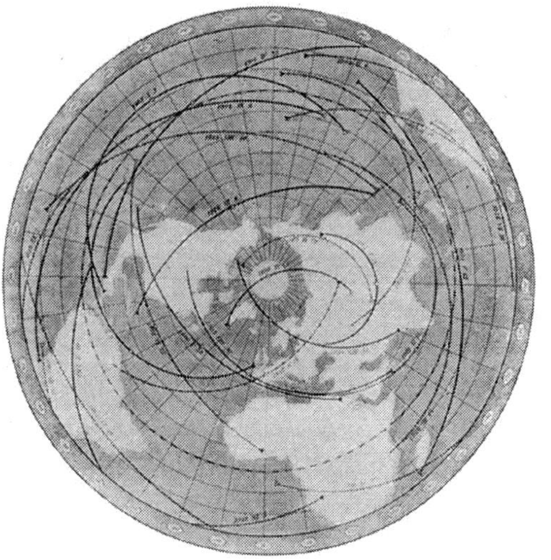

FIG. 36.— Central eclipses for the first two decades of the twentieth century. OPPOLZER.

If we examine the dates of the eclipses shown in this map we shall find that they are not limited to the particular seasons, May and November, in which those of the year 1900 occurred, but are scattered through all the months of the year, from January to December. This shows at once that the line of nodes, *N' N''*, of Fig. 34, does not remain in a fixed position, but turns round in the plane of the earth's orbit so that in different years the earth reaches the node in different months. The precession has already furnished us an illustration of a similar change, the slow rotation of the earth's axis, producing a corresponding shifting of the line in which the planes of the equator and ecliptic intersect; and in much the same way, through the disturbing influence of the sun's attraction, the line *N' N''* is made to revolve westward, opposite to the arrowheads in Fig. 34, at the rate of nearly 20° per year, so that the earth comes to each node about 19 days earlier in each year than in the year preceding, and the eclipse season in each year comes on the average about 19 days earlier than in the year before, although there is a good deal of irregularity in the amount of change in particular years.

72. ***Recurrence of eclipses.***— Before the beginning of the Christian era astronomers had found out a rough-and-ready method of predicting eclipses, which is still of interest and value. The substance of the method is that if we start with any eclipse whatever— e. g., the eclipse of May 28, 1900— and reckon forward or backward from that date a period of 18 years and 10 or 11 days, we shall find another eclipse quite similar in its general characteristics to the one with which we started. Thus, from the map of eclipses (Fig. 36), we find that a total solar eclipse will occur on June 8, 1918, 18 years and 11 days after the one illustrated in Fig. 35. This period of 18 years and 11 days is

called *saros*, an ancient word which means cycle or repetition, and since every eclipse is repeated after the lapse of a saros, we may find the dates of all the eclipses of 1918 by adding 11 days to the dates given in the table of eclipses for 1900 (§ 67), and it is to be especially noted that each eclipse of 1918 will be like its predecessor of 1900 in character– lunar, solar, partial, total, etc. The eclipses of any year may be predicted by a similar reference to those which occurred eighteen years earlier. Consult a file of old almanacs.

The exact length of a saros is 223 lunar months, each of which is a little more than 29.5 days long, and if we multiply the exact value of this last number (see § 60) by 223, we shall find for the product 6,585.32 days, which is equal to 18 years 11.32 days when there are four leap years included in the 18, or 18 years 10.32 days when the number of leap years is five; and in applying the saros to the prediction of eclipses, due heed must be paid to the number of intervening leap years. To explain why eclipses are repeated at the end of the saros, we note that the occurrence of an eclipse depends solely upon the relative positions of the earth, moon, and node of the moon's orbit, and the eclipse will be repeated as often as these three come back to the position which first produced it. This happens at the end of every saros, since the saros is, approximately, the least common multiple of the length of the year, the length of the lunar month, and the length of time required by the line of nodes to make a complete revolution around the ecliptic. If the saros were exactly a multiple of these three periods, every eclipse would be repeated over and over again for thousands of years; but such is not the case, the saros is not an exact multiple of a year, nor is it an exact multiple of the time required for a revolution of the line of nodes, and in consequence the restitution which comes at the end of the saros is not a perfect one. The earth at the 223d new moon is in fact about half a day's motion farther west, relative to the node, than it was at the beginning, and the resulting eclipse, while very similar, is not precisely the same as before. After another 18 years, at the second repetition, the earth is a day farther from the node than at first, and the eclipse differs still more in character, etc. This is shown in Fig. 37, which represents the apparent positions of the disks of the sun and moon as seen from the center of the earth at the end of each sixth saros, 108 years, where the upper row of figures represents the number of repetitions of the eclipse from the beginning, marked *0*, to the end, *72*. The solar eclipse limits, 10, 16, 19 days, are also shown, and all those eclipses which fall between the 10-day limits will be central as seen from some part of the earth, those between 16 and 19 partial wherever seen, while between 10 and 16 they may be either total or partial. Compare the figure with the following description given by Professor Newcomb: "A series of such eclipses commences with a very small eclipse near one pole of the earth. Gradually increasing for about eleven recurrences, it will become central near the same pole. Forty or more central eclipses will then recur, the central line moving slowly toward the other pole. The series will then become partial, and finally cease. The entire duration of the series will be more than a thousand years. A new series commences, on the average, at intervals of thirty years."

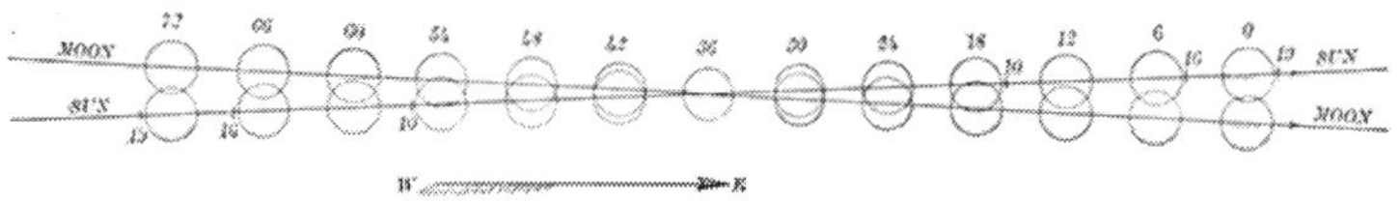

FIG. 37.— Graphical illustration of the saros.

A similar figure may be constructed to represent the recurrence of lunar eclipses; but here, in consequence of the smaller eclipse limits, we shall find that a series is of shorter duration, a little over eight centuries as compared with twelve centuries, which is the average duration of a series of solar eclipses.

One further matter connected with the saros deserves attention. During the period of 6,585.32 days the earth has 6,585 times turned toward the sun the same face upon which the moon's shadow fell at the beginning of the saros, but at the end of the saros the odd 0.32 of a day gives the earth time to make about a third of a revolution more before the eclipse is repeated, and in consequence the eclipse is seen in a different region of the earth, on the average about 116° farther west in longitude. Compare in Fig. 36 the regions in which the eclipses of 1900 and 1918 are visible.

Is this change in the region where the repeated eclipse is visible, true of lunar eclipses as well as solar?

73. ***Use of eclipses.***— At all times and among all peoples eclipses, and particularly total eclipses of the sun, have been reckoned among the most impressive phenomena of Nature. In early times and among uncultivated people they were usually regarded with apprehension, often amounting to a terror and frenzy, which civilized travelers have not scrupled to use for their own purposes with the aid of the eclipse predictions contained in their almanacs, threatening at the proper time to destroy the sun or moon, and pointing to the advancing eclipse as proof that their threats were not vain. In our own day and our own land these feelings of awe have not quite disappeared, but for the most part eclipses are now awaited with an interest and pleasure which, contrasted with the former feelings of mankind, furnish one of the most striking illustrations of the effect of scientific knowledge in transforming human fear and misery into a sense of security and enjoyment.

But to the astronomer an eclipse is more than a beautiful illustration of the working of natural laws; it is in varying degree an opportunity of adding to his store of knowledge respecting the heavenly bodies. The region immediately surrounding the sun is at most times closed to research by the blinding glare of the sun's own light, so that a planet as large as the moon might exist here unseen were it not for the occasional opportunity presented by a total eclipse which shuts off the excessive light and permits not only a search for unknown planets but for anything and everything which may exist around the sun. More than one astronomer has reported the discovery of such planets, and at least one of these has found a name and a description in some of the books, but at the present time most astronomers are very skeptical about the existence of any such object of considerable size, although there is some reason to believe that an enormous number of little bodies, ranging in size from grains of sand upward, do move in this region, as yet unseen and offering to the future problems for investigation.

But in other directions the study of this region at the times of total eclipse has yielded far larger returns, and in the chapter on the sun we shall have to consider the marvelous appearances presented by the solar prominences and by the corona, an appendage of the sun which reaches out from his surface for millions of miles but is never seen save at an eclipse. Photographs of the corona are taken by astronomers at every opportunity, and reproductions of some of these may be found in Chapter X.

Annular eclipses and lunar eclipses are of comparatively little consequence, but any recorded eclipse may become of value in connection with chronology. We date our letters in a particular year of the twentieth century, and commonly suppose that the years are reckoned from the birth of Christ; but this is an error, for the eclipses which were observed of old and by the chroniclers have been associated with events of his life, when examined by the astronomers are found quite inconsistent with astronomic theory. They are, however, reconciled with it if we assume that our system of dates has its origin four years after the birth of Christ, or, in other words, that Christ was born in the year 4 B. C. A mistake was doubtless made at the time the Christian era was introduced into chronology. At many other points the chance record of an eclipse in the early annals of civilization furnishes a similar means of controlling and correcting the dates assigned by the historian to events long past.

CHAPTER VIII.
INSTRUMENTS AND THE PRINCIPLES INVOLVED IN THEIR USE

74. ***Two familiar instruments.***— In previous chapters we have seen that a clock and a divided circle (protractor) are needed for the observations which an astronomer makes, and it is worth while to note here that the geography of the sky and the science of celestial motions depend fundamentally upon these two instruments. The protractor is a simple instrument, a humble member of the family of divided circles, but untold labor and ingenuity have been expended on this family to make possible the construction of a circle so accurately divided that with it angles may be measured to the tenth of a second instead of to the tenth of a degree— i. e., 3,600 times as accurate as the protractor furnishes.

The building of a good clock is equally important and has cost a like amount of labor and pains, so that it is a far cry from Galileo and his discovery that a pendulum "keeps time" to the modern clock with its accurate construction and elaborate provision against disturbing influences of every kind. Every such timepiece, whether it be of the nutmeg variety which sells for a dollar, or whether it be the standard clock of a great national observatory, is made up of the same essential parts that fall naturally into four classes, which we may compare with the departments of a well-ordered factory: I. A timekeeping department, the pendulum or balance spring, whose oscillations must all be of equal duration. II. A power department, the weights or mainspring, which, when wound, store up the power applied from outside and give it out piecemeal as required to keep the first department running. III. A publication department, the dial and hands, which give out the time furnished by Department I. IV. A transportation department, the wheels, which connect the other three and serve as a means of transmitting power and time from one to the other. The case of either clock or watch is merely the roof which shelters it and forms no department of its industry. Of these departments the first is by far the most important, and its good or bad performance makes or mars the credit of the clock. Beware of meddling with the balance wheel of your watch.

75. ***Radiant energy.***— But we have now to consider other instruments which in practice supplement or displace the simple apparatus hitherto employed. Among the most important of these modern instruments are the telescope, the spectroscope, and the photographic camera; and since all these instruments deal with the light which comes from the stars to the earth, we must for their proper understanding take account of the nature of that light, or, more strictly speaking, we must take account of the radiant energy emitted by the sun and stars, which energy, coming from the sun, is translated by our nerves into the two different sensations of light and heat. The radiant energy which comes from the stars is not fundamentally different from that of the

sun, but the amount of energy furnished by any star is so small that it is unable to produce through our nerves any sensible perception of heat, and for the same reason the vast majority of stars are invisible to the unaided eye; they do not furnish a sufficient amount of energy to affect the optic nerves. A hot brick taken into the hand reveals its presence by the two different sensations of heat and pressure (weight); but as there is only one brick to produce the two sensations, so there is only one energy to produce through its action upon different nerves the two sensations of light and heat, and this energy is called *radiant* because it appears to stream forth radially from everything which has the capacity of emitting it. For the detailed study of radiant energy the student is referred to that branch of science called physics; but some of its elementary principles may be learned through the following simple experiment, which the student should not fail to perform for himself:

Drop a bullet or other similar object into a bucket of water and observe the circular waves which spread from the place where it enters the water. These waves are a form of radiant energy, but differing from light or heat in that they are visibly confined to a single plane, the surface of the water, instead of filling the entire surrounding space. By varying the size of the bucket, the depth of the water, the weight of the bullet, etc., different kinds of waves, big and little, may be produced; but every such set of waves may be described and defined in all its principal characteristics by means of three numbers— viz., the vertical height of the waves from hollow to crest; the distance of one wave from the next; and the velocity with which the waves travel across the water. The last of these quantities is called the velocity of propagation; the second is called the wave length; one half of the first is called the amplitude; and all these terms find important applications in the theory of light and heat.

The energy of the falling bullet, the disturbance which it produced on entering the water, was carried by the waves from the center to the edge of the bucket but not beyond, for the wave can go only so far as the water extends. The transfer of energy in this way requires a perfectly continuous medium through which the waves may travel, and the whole visible universe is supposed to be filled with something called *ether*, which serves everywhere as a medium for the transmission of radiant energy just as the water in the experiment served as a medium for transmitting in waves the energy furnished to it by the falling bullet. The student may think of this energy as being transmitted in spherical waves through the ether, every glowing body, such as a star, a candle flame, an arc lamp, a hot coal, etc., being the origin and center of such systems of waves, and determining by its own physical and chemical properties the wave length and amplitude of the wave systems given off.

The intensity of any light depends upon the amplitude of the corresponding vibration, and its color depends upon the wave length. By ingenious devices which need not be here described it has been found possible to measure the wave length corresponding to different colors— e. g., all of the colors of the rainbow, and some of these wave lengths expressed in tenth meters are as follows: A tenth meter is the length obtained by dividing a meter into 10^{10} equal parts. 10^{10} = 10,000,000,000.

Color.			*Wave length.*
Extreme	limit	of visible violet	3,900
Middle	of the	violet	4,060
"	"	blue	4,730
"	"	green	5,270
"	"	yellow	5,810
"	"	orange	5,970
"	"	red	7,000
Extreme	limit	of visible red	7,600

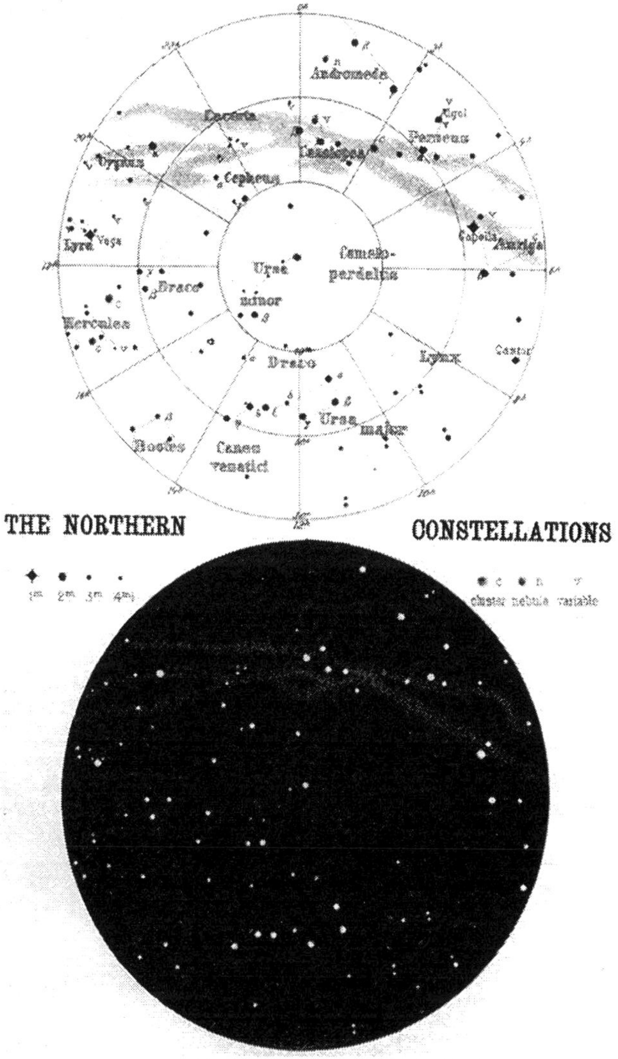

Plate I. The Northern Constellations

The phrase "extreme limit of visible violet" or red used above must be understood to mean that in general the eye is not able to detect radiant energy having a wave length less than 3,900 or greater than 7,600 tenth meters. Radiant energy, however, exists in waves of both greater and shorter length than the above, and may be readily detected by apparatus not subject to the limitations of the human eye— e. g., a common thermometer will show a rise of temperature when its bulb is exposed to radiant energy of wave length much greater than 7,600 tenth meters, and a photographic plate will be strongly affected by energy of shorter wave length than 3,900 tenth meters.

76. ***Reflection and condensation of waves.***— When the waves produced by dropping a bullet into a bucket of water meet the sides of the bucket, they appear to rebound and are reflected back toward the center, and if the bullet is dropped very near the center of the bucket the reflected waves will meet simultaneously at this point and produce there by their combined action a wave higher than that which was reflected at the walls of the bucket. There has been a condensation of energy produced by the reflection, and this increased energy is shown by the greater amplitude of the wave. The student should not fail to notice that each portion of the wave has traveled out and back over the radius of the bucket, and that they meet simultaneously at the center because of this equality of the paths over which they travel, and the resulting equality of time required to go out and back. If the bullet were dropped at one side of the center, would the reflected waves produce *at any point* a condensation of energy?

If the bucket were of elliptical instead of circular cross section and the bullet were dropped at one focus of the ellipse there would be produced a condensation of reflected energy at the other focus, since the sum of the paths traversed by each portion of the wave before and after reflection is equal to the sum of the paths traversed by every other portion, and all parts of the wave reach the second focus at the same time. Upon what geometrical principle does this depend?

The condensation of wave energy in the circular and elliptical buckets are special cases under the general principle that such a condensation will be produced at any point which is so placed that different parts of the wave front reach it simultaneously, whether by reflection or by some other means, as shown below.

The student will note that for the sake of greater precision we here say *wave front* instead of wave. If in any wave we imagine a line drawn along the crest, so as to touch every drop which at that moment is exactly at the crest, we shall have what is called a wave front, and similarly a line drawn through the trough between two waves, or through any set of drops similarly placed on a wave, constitutes a wave front.

77. ***Mirrors and lenses.***— That form of radiant energy which we recognize as light and heat may be reflected and condensed precisely as are the waves of water in the exercise considered above, but owing to the extreme shortness of the wave length in this case the reflecting surface should be very smooth and highly polished. A piece of glass hollowed out in the center by grinding, and with a light film of silver chemically deposited upon the hollow

surface and carefully polished, is often used by astronomers for this purpose, and is called a concave mirror.

The radiant energy coming from a star or other distant object and falling upon the silvered face of such a mirror is reflected and condensed at a point a little in front of the mirror, and there forms an image of the star, which may be seen with the unaided eye, if it is held in the right place, or may be examined through a magnifying glass. Similarly, an image of the sun, a planet, or a distant terrestrial object is formed by the mirror, which condenses at its appropriate place the radiant energy proceeding from each and every point in the surface of the object, and this, in common phrase, produces an image of the object.

Another device more frequently used by astronomers for the production of images (condensation of energy) is a lens which in its simplest form is a round piece of glass, thick in the center and thin at the edge, with a cross section, such as is shown at *A B* in Fig. 38. If we suppose *E G D* to represent a small part of a wave front coming from a very distant source of radiant energy, such as a star, this wave front will be practically a plane surface represented by the straight line *E D*, but in passing through the lens this surface will become warped, since light travels slower in glass than in air, and the central part of the beam, *G*, in its onward motion will be retarded by the thick center of the lens, more than *E* or *D* will be retarded by the comparatively thin outer edges of *A B*. On the right of the lens the wave front therefore will be transformed into a curved surface whose exact character depends upon the shape of the lens and the kind of glass of which it is made. By properly choosing these the new wave front may be made a part of a sphere having its center at the point *F* and the whole energy of the wave front, *E G D*, will then be condensed at *F*, because this point is equally distant from all parts of the warped wave front, and therefore is in a position to receive them simultaneously. The distance of *F* from *A B* is called the focal length of the lens, and *F* itself is called the focus. The significance of this last word (Latin, *focus* = fireplace) will become painfully apparent to the student if he will hold a common reading glass between his hand and the sun in such a way that the focus falls upon his hand.

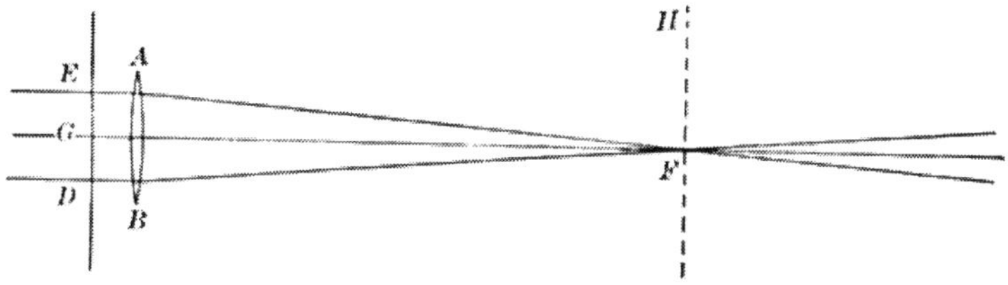

FIG. 38.— Illustrating the theory of lenses.

All the energy transmitted by the lens in the direction *G F* is concentrated upon a very small area at *F*, and an image of the object– e. g., a star, from which the light came– is formed here. Other stars situated near the one in question will also send beams of light along slightly different directions to the lens, and these will be concentrated, each in its appropriate place, in the *focal plane*, *F H*, passed through the focus, *F*, perpendicular to the line, *F G*, and we shall find in this plane a picture of all the stars or other objects within the range of the lens.

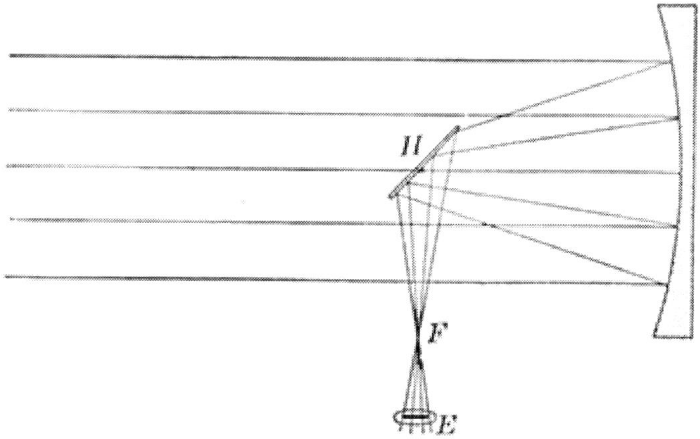

*FIG. 39.— **Essential parts of a reflecting telescope.***

78. ***Telescopes.***— The simplest kind of telescope consists of a concave mirror to produce images, and a magnifying glass, called an *eyepiece*, through which to examine them; but for convenience' sake, so that the observer may not stand in his own light, a small mirror is frequently added to this combination, as at *H* in Fig. 39, where the lines represent the directions along which the energy is propagated. By reflection from this mirror the focal plane and the images are shifted to *F*, where they may be examined from one side through the magnifying glass *E*.

Such a combination of parts is called a *reflecting* telescope, while one in which the images are produced by a lens or combination of lenses is called a *refracting* telescope, the adjective having reference to the bending, refraction, produced by the glass upon the direction in which the energy is propagated. The customary arrangement of parts in such a telescope is shown in Fig. 40, where the part marked *O* is called the objective and *V E* (the magnifying glass) is the eyepiece, or ocular, as it is sometimes called.

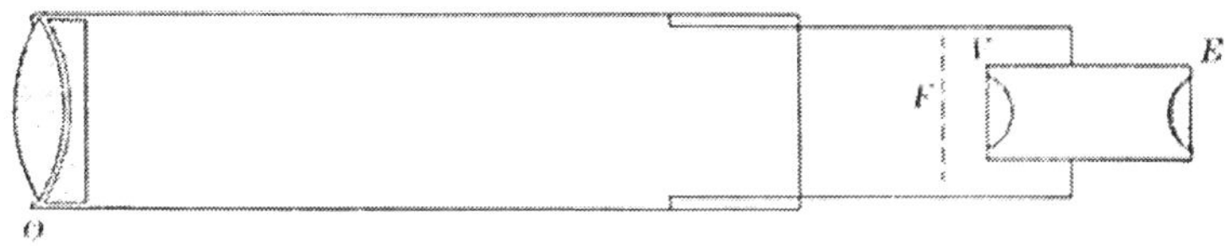

FIG. 40.—A simple form of refracting telescope.

Most objects with which we have to deal in using a telescope send to it not light of one color only, but a mixture of light of many colors, many different wave lengths, some of which are refracted more than others by the glass of which the lens is composed, and in consequence of these different amounts of refraction a single lens does not furnish a single image of a star, but gives a confused jumble of red and yellow and blue images much inferior in sharpness of outline (definition) to the images made by a good concave mirror. To remedy this defect it is customary to make the objective of two or more pieces of glass of different densities and ground to different shapes as is shown at *O* in Fig. 40. The two pieces of glass thus mounted in one frame constitute a compound lens having its own focal plane, shown at *F* in the figure, and similarly the lenses composing the eyepiece have a focal plane between the eyepiece and the objective which must also fall at *F*, and in the use of a telescope the eyepiece must be pushed out or in until its focal plane coincides with that of the objective. This process, which is called focusing, is what is accomplished in the ordinary opera glass by turning a screw placed between the two tubes, and it must be carefully done with every telescope in order to obtain distinct vision.

79. ***Magnifying power.***— The amount by which a given telescope magnifies depends upon the focal length of the objective (or mirror) and the focal length of the eyepiece, and is equal to the ratio of these two quantities. Thus in Fig. 40 the distance of the objective from the focal plane *F* is about 16 times as great as the distance of the eyepiece from the same plane, and the magnifying power of this telescope is therefore 16 diameters. A magnifying power of 16 diameters means that the diameter of any object seen in the telescope looks 16 times as large as it appears without the telescope, and is nearly equivalent to saying that the object appears only one sixteenth as far off. Sometimes the magnifying power is assumed to be the number of times that the *area* of an object seems increased; and since areas are proportional to the squares of lines, the magnifying power of 16 diameters might be called a power of 256. Every large telescope is provided with several eyepieces of different focal lengths, ranging from a quarter of an inch to two and a half inches, which are used to furnish different magnifying powers as may be required for the different kinds of work undertaken with the instrument. Higher powers can be used with large telescopes than with small ones, but it

is seldom advantageous to use with any telescope an eyepiece giving a higher power than 60 diameters for each inch of diameter of the objective.

The part played by the eyepiece in determining magnifying power will be readily understood from the following experiment:

Make a pin hole in a piece of cardboard. Bring a printed page so close to one eye that you can no longer see the letters distinctly, and then place the pin hole between the eye and the page. The letters which were before blurred may now be seen plainly through the pin hole, even when the page is brought nearer to the eye than before. As it is brought nearer, notice how the letters seem to become larger, solely because they are nearer. A pin hole is the simplest kind of a magnifier, and the eyepiece in a telescope plays the same part as does the pin hole in the experiment; it enables the eye to be brought nearer to the image, and the shorter the focal length of the eyepiece the nearer is the eye brought to the image and the higher is the magnifying power.

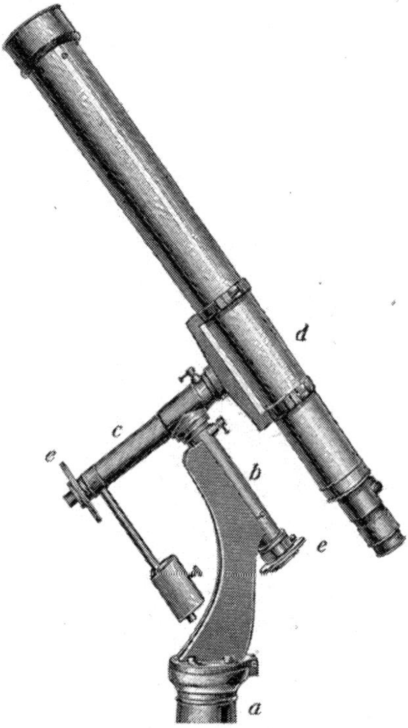

FIG. 41.—A simple equatorial mounting.

80. ***The equatorial mounting.***— Telescopes are of all sizes, from the modest opera glass which may be carried in the pocket and which requires no other support than the hand, to the giant which must have a special roof to shelter it and elaborate machinery to support and direct it toward the sky. But

for even the largest telescopes this machinery consists of the following parts, which are illustrated, with exception of the last one, in the small equatorial telescope shown in Fig. 41. It is not customary to place a driving clock on so small a telescope as this:

(*a*) A supporting pier or tripod.

(*b*) An axis placed parallel to the axis of the earth.

(*c*) Another axis at right angles to *b* and capable of revolving upon *b* as an axle.

(*d*) The telescope tube attached to *c* and capable of revolving about *c*.

(*e*) Graduated circles attached to *c* and *b* to measure the amount by which the telescope is turned on these axes.

(*f*) A driving clock so connected with *b* as to make *c* (and *d*) revolve about *b* with an angular velocity equal and opposite to that with which the earth turns upon its axis.

Such a support is called an equatorial mounting, and the student should note from the figure that the circles, *e*, measure the hour angle and declination of any star toward which the telescope is directed, and conversely if the telescope be so set that these circles indicate the hour angle and declination of any given star, the telescope will then point toward that star. In this way it is easy to find with the telescope any moderately bright star, even in broad daylight, although it is then absolutely invisible to the naked eye. The rotation of the earth about its axis will speedily carry the telescope away from the star, but if the driving clock be started, its effect is to turn the telescope toward the west just as fast as the earth's rotation carries it toward the east, and by these compensating motions to keep it directed toward the star. In Fig. 42, which represents the largest and one of the most perfect refracting telescopes ever built, let the student pick out and identify the several parts of the mounting above described. A part of the driving clock may be seen within the head of the pier. In Fig. 43 trace out the corresponding parts in the mounting of a reflecting telescope.

FIG. 42.— Equatorial mounting of the great telescope of the Yerkes Observatory.

FIG. 43.— The reflecting telescope of the Paris Observatory.

A telescope is often only a subordinate part of some instrument or apparatus, and then its style of mounting is determined by the requirements of the special case; but when the telescope is the chief thing, and the remainder of the apparatus is subordinate to it, the equatorial mounting is almost always adopted, although sometimes the arrangement of the parts is very different in appearance from any of those shown above. Beware of the popular error that an object held close in front of a telescope can be seen by an observer at the eyepiece. The numerous stories of astronomers who saw spiders crawling over the objective of their telescope, and imagined they were beholding strange objects in the sky, are all fictitious, since nothing on or near the objective could possibly be seen through the telescope.

81. ***Photography.*** – A photographic camera consists of a lens and a device for holding at its focus a specially prepared plate or film. This plate carries a chemical deposit which is very sensitive to the action of light, and which may be made to preserve the imprint of any picture which the lens forms upon it. If such a sensitive plate is placed at the focus of a reflecting telescope, the combination becomes a camera available for astronomical photography, and at the present time the tendency is strong in nearly every branch of astronomical research to substitute the sensitive plate in place of the observer at a telescope. A refracting telescope may also be used for astronomical photography, and is very much used, but some complications occur here on account of the resolution of the light into its constituent colors in passing through the objective. Fig. 44 shows such a telescope, or rather two telescopes, one photographic, the other visual, supported side by side upon the same equatorial mounting.

*FIG. 44.— **Photographic telescope of the Paris Observatory.***

One of the great advantages of photography is found in connection with what is called–

82. ***Personal equation.***– It is a remarkable fact, first investigated by the German astronomer Bessel, three quarters of a century ago, that where extreme accuracy is required the human senses can not be implicitly relied upon. The most skillful observers will not agree exactly in their measurement of an angle or in estimating the exact instant at which a star crossed the meridian; the most skillful artists can not draw identical pictures of the same object, etc.

These minor deceptions of the senses are included in the term *personal equation*, which is a famous phrase in astronomy, denoting that the observations of any given person require to be corrected by means of some equation involving his personality.

General health, digestion, nerves, fatigue, all influence the personal equation, and it was in reference to such matters that one of the most eminent of living astronomers has given this description of his habits of observing:

"In order to avoid every physiological disturbance, I have adopted the rule to abstain for one or two hours before commencing observations from every laborious occupation; never to go to the telescope with stomach loaded with food; to abstain from everything which could affect the nervous system, from narcotics and alcohol, and especially from the abuse of coffee, which I have found to be exceedingly prejudicial to the accuracy of observation."[C] A regimen suggestive of preparation for an athletic contest rather than for the more quiet labors of an astronomer.

83. ***Visual and photographic work.***— The photographic plate has no stomach and no nerves, and is thus free from many of the sources of error which inhere in visual observations, and in special classes of work it possesses other marked advantages, such as rapidity when many stars are to be dealt with simultaneously, permanence of record, and owing to the cumulative effect of long exposure of the plate it is possible to photograph with a given telescope stars far too faint to be seen through it. On the other hand, the eye has the advantage in some respects, such as studying the minute details of a fairly bright object— e. g., the surface of a planet, or the sun's corona and, for the present at least, neither method of observing can exclude the other. For a remarkable case of discordance between the results of photographic and visual observations compare the pictures of the great nebula in the constellation Andromeda, which are given in Chapter XIV. A partial explanation of these discordances and other similar ones is that the eye is most strongly affected by greenish-yellow light, while the photographic plate responds most strongly to violet light; the photograph, therefore, represents things which the eye has little capacity for seeing, and *vice versa*.

84. ***The spectroscope.***— In some respects the spectroscope is the exact counterpart of the telescope. The latter condenses radiant energy and the former disperses it. As a measuring instrument the telescope is mainly concerned with the direction from which light comes, and the different colors of which that light is composed affect it only as an obstacle to be overcome in its construction. On the other hand, with the spectroscope the direction from which the radiant energy comes is of minor consequence, and the all-important consideration is the intrinsic character of that radiation. What colors are present in the light and in what proportions? What can these colors be made to tell about the nature and condition of the body from which they come, be it sun, or star, or some terrestrial source of light, such as an arc lamp, a candle flame, or a furnace in blast? These are some of the characteristic questions of the spectrum analysis, and, as the name implies, they are solved by analyzing the radiant energy into its component parts, setting down the blue light in one place, the yellow in another, the red in still another, etc., and interpreting this array of colors by means of principles which we shall have to consider. Something of this process of color analysis may be seen in the brilliant hues shown by a soap bubble, or reflected from a piece of mother-of-pearl, and still more strikingly exhibited in the rainbow,

produced by raindrops which break up the sunlight into its component colors and arrange them each in its appropriate place. Any of these natural methods of decomposing light might be employed in the construction of a spectroscope, but in spectroscopes which are used for analyzing the light from feeble sources, such as a star, or a candle flame, a glass prism of triangular cross section is usually employed to resolve the light into its component colors, which it does by refracting it as shown at the edges of the lens in Fig. 38.

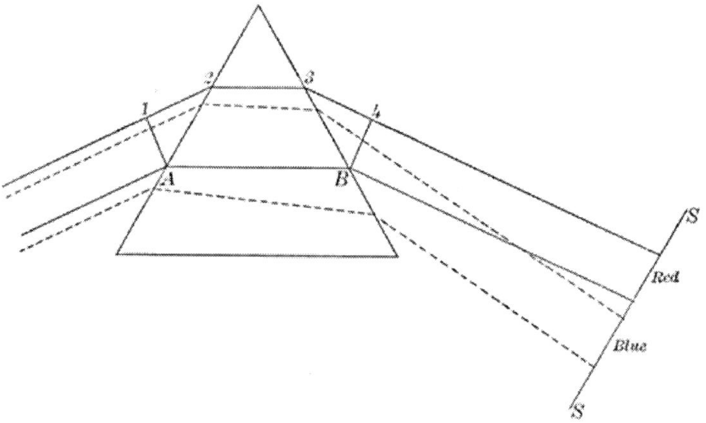

FIG. 45.— ***Resolution of light into its component colors.***

The course of a beam of light in passing through such a prism is shown in Fig. 45. Note that the bending of the light from its original course into a new one, which is here shown as produced by the prism, is quite similar to the bending shown at the edges of a lens and comes from the same cause, the slower velocity of light in glass than in air. It takes the light-waves as long to move over the path *A B* in glass as over the longer path *1*, *2*, *3*, *4*, of which only the middle section lies in the glass.

Not only does the prism bend the beam of light transmitted by it, but it bends in different degree light of different colors, as is shown in the figure, where the beam at the left of the prism is supposed to be made up of a mixture of blue and red light, while at the right of the prism the greater deviation imparted to the blue quite separates the colors, so that they fall at different places on the screen, *S S*. The compound light has been analyzed into its constituents, and in the same way every other color would be put down at its appropriate place on the screen, and a beam of white light falling upon the prism would be resolved by it into a sequence of colors, falling upon the screen in the order red, orange, yellow, green, blue, indigo, violet. The initial letters of these names make the word *Roygbiv*, and by means of it their order is easily remembered.

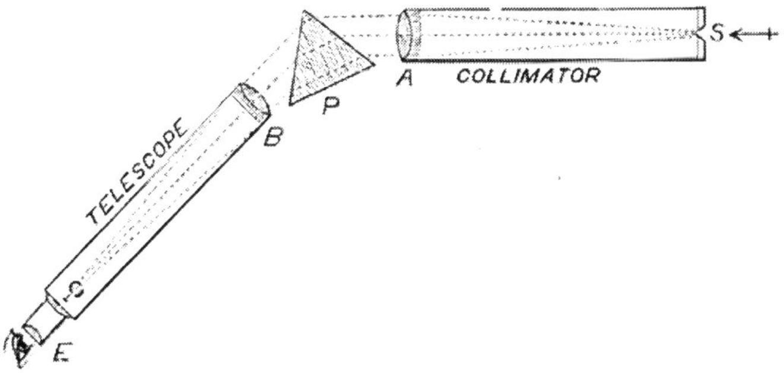

FIG. 46.— ***Principal parts of a spectroscope.***

If the light which is to be examined comes from a star the analysis made by the prism is complete, and when viewed through a telescope the image of the star is seen to be drawn out into a band of light, which is called a *spectrum*, and is red at one end and violet or blue at the other, with all the colors of the rainbow intervening in proper order between these extremes. Such a prism placed in front of the objective of a telescope is called an objective prism, and has been used for stellar work with marked success at the Harvard College Observatory. But if the light to be analyzed comes from an object having an appreciable extent of surface, such as the sun or a planet, the objective prism can not be successfully employed, since each point of the surface will produce its own spectrum, and these will appear in the *view telescope* superposed and confused one with another in a very objectionable manner. To avoid this difficulty there is placed between the prism and the source of light an opaque screen, *S*, with a very narrow slit cut in it, through which all the light to be analyzed must pass and must also go through a lens, *A*, placed between the slit and the prism, as shown in Fig. 46. The slit and lens, together with the tube in which they are usually supported, are called a *collimator*. By this device a very limited amount of light is permitted to pass from the object through the slit and lens to the prism and is there resolved into a spectrum, which is in effect a series of images of the slit in light of different colors, placed side by side so close as to make practically a continuous ribbon of light whose width is the length of each individual picture of the slit. The length of the ribbon (dispersion) depends mainly upon the shape of the prism and the kind of glass of which it is made, and it may be very greatly increased and the efficiency of the spectroscope enhanced by putting two, three, or more prisms in place of the single one above described. When the amount of light is very great, as in the case of the sun or an electric arc lamp, it is advantageous to alter slightly the arrangement of the spectroscope and to substitute in place of the prism a grating— i. e., a metallic mirror with a great number of fine parallel lines ruled upon its surface at equal intervals, one from another. It is by virtue of such a system of fine parallel grooves that mother-of-pearl displays its beautiful color effects, and a brilliant spectrum of great purity and high dispersion is furnished by a grating ruled with from 10,000 to 20,000 lines to the inch. Fig. 47 represents, rather crudely, a part of the spectrum of an arc light furnished by such a grating, or rather it shows three different spectra arranged side by side, and looking something like a rude ladder. The sides of the ladder are the spectra furnished by the incandescent carbons of the lamp, and the cross pieces are the spectrum of the electric arc filling the space between the carbons. Fig. 48 shows a continuation of the same spectra into a region where the radiant energy is invisible to the eye, but is capable of being photographed.

*FIG. 47.— **Green and blue part of the spectrum of an electric arc light.***

It is only when a lens is placed between the lamp and the slit of the spectroscope that the three spectra are shown distinct from each other as in the figure. The purpose of the lens is to make a picture of the lamp upon the slit, so that all the radiant energy from any one point of the arc may be brought to one part of the slit, and thus appear in the resulting spectrum separated from the energy which comes from every other part of the arc. Such an instrument is called an *analyzing spectroscope* while one without the lens is called an *integrating spectroscope*, since it furnishes to each point of the slit a sample of the radiant energy coming from every part of the source of light, and thus produces only an average spectrum of that source without distinction of its parts. When a spectroscope is attached to a telescope, as is often done (see Fig. 49), the eyepiece is removed to make way for it, and the telescope objective takes the part of the analyzing lens. A camera is frequently combined with such an apparatus to photograph the spectra it furnishes, and the whole instrument is then called a *spectrograph*.

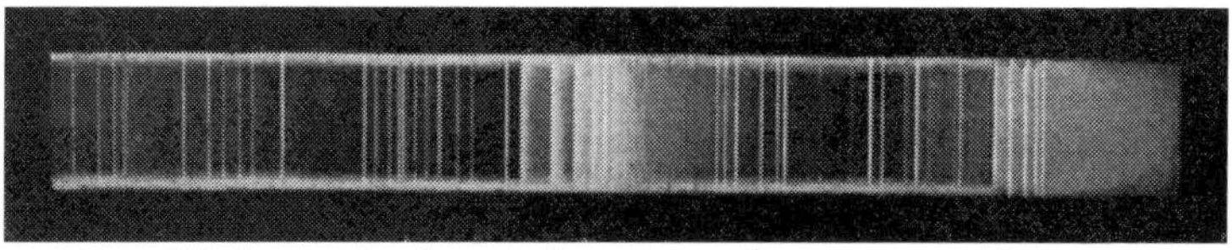

*FIG. 48.— **Violet and ultraviolet parts of spectrum of an arc lamp.***

85. ***Spectrum analysis.***— Having seen the mechanism of the spectroscope by which the light incident upon it is resolved into its constituent parts and drawn out into a series of colors arranged in the order

of their wave lengths, we have now to consider the interpretation which is to be placed upon the various kinds of spectra which may be seen, and here we rely upon the experience of physicists and chemists, from whom we learn as follows:

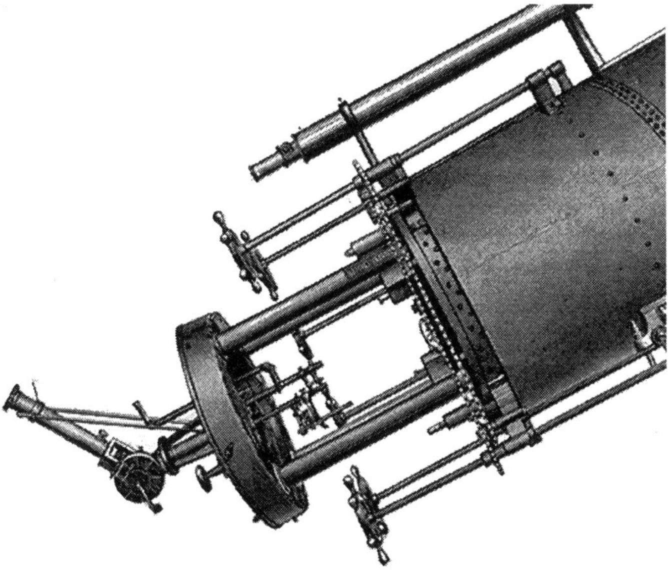

FIG. 49.—A spectroscope attached to the Yerkes telescope.

The radiant energy which is analyzed by the spectroscope has its source in the atoms and molecules which make up the luminous body from which the energy is radiated, and these atoms and molecules are able to impress upon the ether their own peculiarities in the shape of waves of different length and amplitude. We have seen that by varying the conditions of the experiment different kinds of waves may be produced in a bucket of water; and as a study of these waves might furnish an index to the conditions which produced them, so the study of the waves peculiar to the light which comes from any source may be made to give information about the molecules which make up that source. Thus the molecules of iron produce a system of waves peculiar to themselves and which can be duplicated by nothing else, and every other substance gives off its own peculiar type of energy, presenting a limited and definite number of wave lengths dependent upon the nature and condition of its molecules. If these molecules are free to behave in their own characteristic fashion without disturbance or crowding, they emit light of these wave lengths only, and we find in the spectrum a series of bright lines, pictures of the slit produced by light of these particular wave lengths, while between these bright lines lie dark spaces showing the absence from the radiant energy of light of intermediate wave lengths. Such a spectrum is shown in the central portion of Fig. 47, which, as we have already seen, is produced by the space between the carbons of the arc lamp. On the other hand, if the molecules are closely packed together under pressure they so interfere with each other as to give off a jumble of energy of all wave lengths, and this is translated by the spectroscope into a continuous ribbon of light with no dark spaces intervening, as in the upper and lower parts of Figs. 47 and 48, produced by the incandescent solid carbons of the lamp. These two types are known as the continuous and discontinuous spectrum, and we may lay down the following principle regarding them:

A discontinuous spectrum, or bright-line spectrum as it is familiarly called, indicates that the molecules of the source of light are not crowded together, and therefore the light must come from an incandescent gas. A continuous spectrum shows only that the molecules are crowded together, or are so numerous that the body to which they belong is not transparent and gives no further information. The body may be solid, liquid, or gaseous, but in the latter case the gas must be under considerable pressure or of great extent.

A second principle is: The lines which appear in a spectrum are characteristic of the source from which the light came– e. g., the double line in the yellow part of the spectrum at the extreme left in Fig. 47 is produced by sodium vapor in and around the electric arc and is never produced by anything but sodium. When by laboratory experiments we have learned the particular set of lines corresponding to iron, we may treat the presence of these lines in another spectrum as proof that iron is present in the source from which the light came, whether that source be a white-hot poker in the next room or a star immeasurably distant. The evidence that iron is present lies in the nature of the light, and there is no reason to suppose that nature to be altered on the way from star to earth. It may, however, be altered by something happening to the source from which it comes– e. g., changing temperature or pressure may affect, and does affect, the spectrum which such a substance as iron emits, and we must be prepared to find the same substance presenting different spectra under different conditions, only these conditions must be greatly altered in order to produce radical changes in the spectrum.

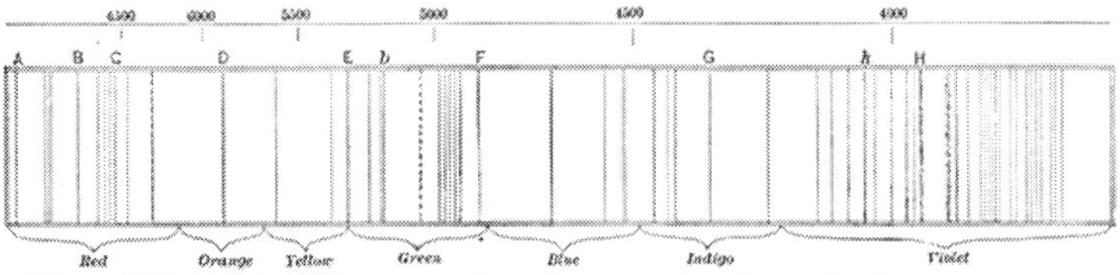

FIG. 50.— The chief lines in the spectrum of sunlight.— HERSCHEL.

86. ***Wave lengths.***– To identify a line as belonging to and produced by iron or any other substance, its position in the spectrum– i. e., its wave

length– must be very accurately determined, and for the identification of a substance by means of its spectrum it is often necessary to determine accurately the wave lengths of many lines. A complicated spectrum may consist of hundreds or thousands of lines, due to the presence of many different substances in the source of light, and unless great care is taken in assigning the exact position of these lines in the spectrum, confusion and wrong identifications are sure to result. For the measurement of the required wave length a tenth meter (§ 75) is the unit employed, and a scale of wave lengths expressed in this unit is presented in Fig. 50. The accuracy with which some of these wave lengths are determined is truly astounding; a ten-billionth of an inch! These numerical wave lengths save all necessity for referring to the color of any part of the spectrum, and pictures of spectra for scientific use are not usually printed in colors.

87. ***Absorption spectra.*** – There is another kind of spectrum, of greater importance than either of those above considered, which is well illustrated by the spectrum of sunlight (Fig. 50). This is a nearly continuous spectrum crossed by numerous *dark* lines due to absorption of radiant energy in a comparatively cool gas through which it passes on its way to the spectroscope. Fraunhofer, who made the first careful study of spectra, designated some of the more conspicuous of these lines by letters of the alphabet which are shown in the plate, and which are still in common use as names for the lines, not only in the spectrum of sunlight but wherever they occur in other spectra. Thus the double line marked *D*, wave length 5893, falls at precisely the same place in the spectrum as does the double (sodium) line which we have already seen in the yellow part of the arc-light spectrum, which line is also called *D* and bears a very intimate relation to the dark *D* line of the solar spectrum.

The student who has access to colored crayons should color one edge of Fig. 50 in accordance with the lettering there given and, so far as possible, he should make the transition from one color to the next a gradual one, as it is in the rainbow.

Fig. 50 is far from being a complete representation of the spectrum of sunlight. Not only does this spectrum extend both to the right and to the left into regions invisible to the human eye, but within the limits of the figure, instead of the seventy-five lines there shown, there are literally thousands upon thousands of lines, of which only the most conspicuous can be shown in such a cut as this.

The dark lines which appear in the spectrum of sunlight can, under proper conditions, be made to appear in the spectrum of an arc light, and Fig. 51 shows a magnified representation of a small part of such a spectrum adjacent to the *D* (sodium) lines. Down the middle of each of these lines runs a black streak whose position (wave length) is precisely that of the *D* lines in the spectrum of sunlight, and whose presence is explained as follows:

The very hot sodium vapor at the center of the arc gives off its characteristic light, which, shining through the outer and cooler layers of sodium vapor, is partially absorbed by these, resulting in a fine dark line corresponding exactly in position and wave length to the bright lines, and seen against these as a background, since the higher temperature at the center of the arc tends to broaden the bright lines and make them diffuse. Similarly the dark lines in the spectrum of the sun (Fig. 50) point to the existence of a surrounding envelope of relatively cool gases, which absorb from the sunlight precisely those kinds of radiant energy which they would themselves emit if incandescent. The resulting dark lines in the spectrum are to be interpreted by the same set of principles which we have above applied to the bright lines of a discontinuous spectrum, and they may be used to determine the chemical composition of the sun, just as the bright lines serve to determine the chemical elements present in the electric arc. With reference to the mode of their formation, bright-line and dark-line spectra are sometimes called respectively *emission* and *absorption* spectra.

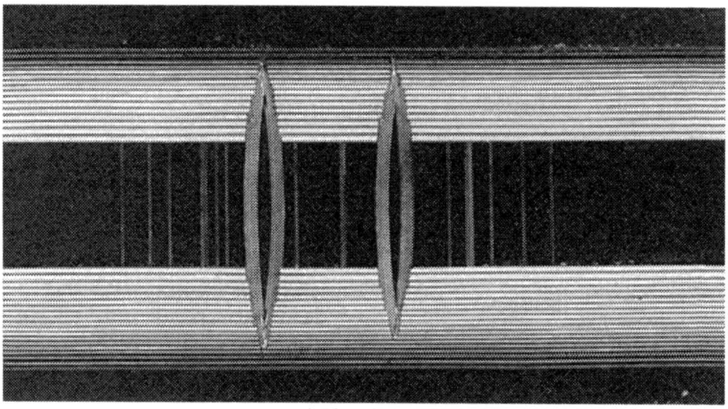

FIG. 51.— The lines reversed.

88. ***Types of spectrum.***— The sun presents by far the most complex spectrum known, and Fig. 50 shows only a small number of the more conspicuous lines which appear in it. Spectra of stars, *per contra*, appear relatively simple, since their feeble light is insufficient to bring out faint details. In Chapters XIII and XIV there are shown types of the different kinds of spectra given by starlight, and these are to be interpreted by the principles above established. Thus the spectrum of the bright star β Aurigæ shows a continuous spectrum crossed by a few heavy absorption lines which are known from laboratory experiments to be produced only by hydrogen. There must therefore be an atmosphere of relatively cool hydrogen surrounding this star. The spectrum of Pollux is quite similar to that of the

sun and is to be interpreted as showing a physical condition similar to that of the sun, while the spectrum of α Herculis is quite different from either of the others. In subsequent chapters we shall have occasion to consider more fully these different types of spectrum.

89. ***The Doppler principle.*** – This important principle of the spectrum analysis is most readily appreciated through the following experiment:

Listen to the whistle of a locomotive rapidly approaching, and observe how the pitch changes and the note becomes more grave as the locomotive passes by and commences to recede. During the approach of the whistle each successive sound wave has a shorter distance to travel in coming to the ear of the listener than had its predecessor, and in consequence the waves appear to come in quicker succession, producing a higher note with a correspondingly shorter wave length than would be heard if the same whistle were blown with the locomotive at rest. On the other hand, the wave length is increased and the pitch of the note lowered by the receding motion of the whistle. A similar effect is produced upon the wave length of light by a rapid change of distance between the source from which it comes and the instrument which receives it, so that a diminishing distance diminishes very slightly the wave length of every line in the spectrum produced by the light, and an increasing distance increases these wave lengths, and this holds true whether the change of distance is produced by motion of the source of light or by motion of the instrument which receives it.

This change of wave length is sometimes described by saying that when a body is rapidly approaching, the lines of its spectrum are all displaced toward the violet end of the spectrum, and are correspondingly displaced toward the red end by a receding motion. The amount of this shifting, when it can be measured, measures the velocity of the body along the line of sight, but the observations are exceedingly delicate, and it is only in recent years that it has been found possible to make them with precision. For this purpose there is made to pass through the spectroscope light from an artificial source which contains one or more chemical elements known to be present in the star which is to be observed, and the corresponding lines in the spectrum of this light and in the spectrum of the star are examined to determine whether they exactly match in position, or show, as they sometimes do, a slight displacement, as if one spectrum had been slipped past the other. The difficulty of the observations lies in the extremely small amount of this slipping, which rarely if ever in the case of a moving star amounts to one sixth part of the interval between the close parallel lines marked *D* in Fig. 50. The spectral lines furnished by the headlight of a locomotive running at the rate of a hundred miles per hour would be displaced by this motion less than one six-thousandth part of the space between the *D* lines, an amount absolutely imperceptible in the most powerful spectroscope yet constructed. But many of the celestial bodies have velocities so much greater than a hundred miles per hour that these may be detected and measured by means of the Doppler principle.

90. ***Other instruments.***— Other instruments of importance to the astronomer, but of which only casual mention can here be made, are the meridian-circle; the transit, one form of which is shown in Fig. 52, and the zenith telescope, which furnish refined methods for making observations similar in kind to those which the student has already learned to make with plumb line and protractor; the sextant, which is pre-eminently the sailor's instrument for finding the latitude and longitude at sea, by measuring the altitudes of sun and stars above the sea horizon; the heliometer, which serves for the very accurate measurement of small angles, such as the angular distance between two stars not more than one or two degrees apart; and the photometer, which is used for measuring the amount of light received from the celestial bodies.

*FIG. 52.—**A combined transit instrument and zenith telescope.***

CHAPTER IX.
THE MOON

91. ***Results of observation with the unaided eye.*** – The student who has made the observations of the moon which are indicated in Chapter III has in hand data from which much may be learned about the earth's satellite. Perhaps the most striking feature brought out by them is the motion of the moon among the stars, always from west toward east, accompanied by that endless series of changes in shape and brightness– new moon, first quarter, full moon, etc.– whose successive stages we represent by the words, the phase of the moon. From his own observation the student should be able to verify, at least approximately, the following statements, although the degree of numerical precision contained in some of them can be reached only by more elaborate apparatus and longer study than he has given to the subject:

A. The phase of the moon depends upon the distance apart of sun and moon in the sky, new moon coming when they are together, and full moon when they are as far apart as possible.

The Moon, One Day After First Quarter. From a photograph made at the Paris Observatory.

B. The moon is essentially a round, dark body, giving off no light of its own, but shining solely by reflected sunlight. The proof of this is that whenever we see a part of the moon which is turned away from the sun it looks dark– e. g., at new moon, sun and moon are in nearly the same direction from us and we see little or nothing of the moon, since the side upon which the sun shines is turned away from us. At full moon the earth is in line between sun and moon, and we see, round and bright, the face upon which the sun shines. At other phases, such as the quarters, the moon turns toward the earth a part of its night hemisphere and a part of its day hemisphere, but in general only that part which belongs to the day side of the moon is visible and the peculiar curved line which forms the boundary— the "ragged edge," or *terminator*, as it is called, is the dividing line between day and night upon the moon.

A partial exception to what precedes is found for a few days after new moon when the moon and sun are not very far apart in the sky, for then the whole round disk of the moon may often be seen, a small part of it brightly illuminated by the sun and the larger part feebly illuminated by sunlight which fell first upon the earth and was by it reflected back to the moon, giving the pleasing effect which is sometimes called the old moon in the new moon's arms. The new moon— i. e., the part illumined by the sun— usually appears to belong to a sphere of larger radius than the old moon, but this is purely a trick played by the eyes of the observer, and the effect disappears altogether in a telescope. Is there any similar effect in the few days before new moon?

C. The moon makes the circuit of the sky from a given star around to the same star again in a little more than 27 days (27.32166), but the interval between successive new moons— i. e., from the sun around to the sun again— is more than 29 days (29.53059). This last interval, which is called a lunar month or *synodical* month, indicates what we have learned before— that the sun has changed its place among the stars during the month, so that it takes the moon an extra two days to overtake him after having made the circuit of the sky, just as it takes the minute hand of a clock an extra 5 minutes to catch up with the hour hand after having made a complete circuit of the dial.

D. Wherever the moon may be in the sky, it turns always the same face toward the earth, as is shown by the fact that the dark markings which appear on its surface stand always upon (nearly) the same part of its disk. It does not always turn the same face toward the sun, for the boundary line between the illumined and unillumined parts of the moon shifts from one side to the other as the phase changes, dividing at each moment day from night upon the moon and illustrating by its slow progress that upon the moon the day and the month are of equal length (29.5 terrestrial days), instead of being time units of different lengths as with us.

92. ***The moon's motion.***— The student should compare the results of his own observations, as well as the preceding section, with Fig. 53, in which the lines with dates printed on them are all supposed to radiate from the sun and to represent the direction from the sun of earth and moon upon the given dates which are arbitrarily assumed for the sake of illustration, any other set would do equally well. The black dots, small and large, represent the moon revolving about the earth, but having the circular path shown in Fig. 34

(ellipse) transformed by the earth's forward motion into the peculiar sinuous line here shown. With respect to both earth and sun, the moon's orbit deviates but little from a circle, since the sinuous curve of Fig. 53 follows very closely the earth's orbit around the sun and is almost identical with it.

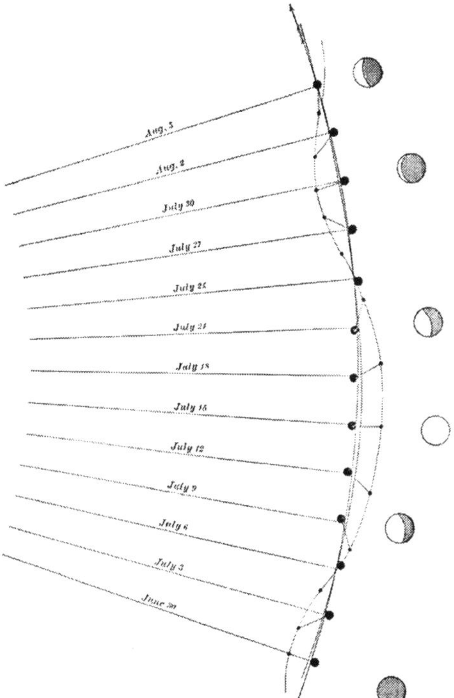

FIG. 53.— Motion of moon and earth relative to the sun.

For clearness of representation the distance between earth and moon in the figure has been made ten times too great, and to get a proper idea of the moon's orbit with reference to the sun, we must suppose the moon moved up toward the earth until its distance from the line of the earth's orbit is only a tenth part of what it is in the figure. When this is done, the moon's path becomes almost indistinguishable from that of the earth, as may be seen in the figure, where the attempt has been made to show both lines, and it is to be especially noted that this real orbit of the moon is everywhere concave toward the sun.

The phase presented by the moon at different parts of its path is indicated by the row of circles at the right, and the student should show why a new moon is associated with June 30th and a full moon with July 15th, etc. What was the date of first quarter? Third quarter?

We may find in Fig. 53 another effect of the same kind as that noted above in C. Between noon, June 30th, and noon, July 3d, the earth makes upon its axis three complete revolutions with respect to the sun, but the meridian which points toward the moon at noon on June 30th will not point toward it at noon on July 3d, since the moon has moved into a new position and is now 37° away from the meridian. Verify this statement by measuring, in Fig. 53, with the protractor, the moon's angular distance from the meridian at noon on July 3d. When will the meridian overtake the moon?

93. ***Harvest moon.***— The interval between two successive transits of the meridian past the moon is called a lunar day, and the student should show from the figure that on the average a lunar day is 51 minutes longer than a solar day— i. e., upon the average each day the moon comes to the meridian 51 minutes of solar time later than on the day before. It is also true that on the average the moon rises and sets 51 minutes later each day than on the day before. But there is a good deal of irregularity in the retardation of the time of moonrise and moonset, since the time of rising depends largely upon the particular point of the horizon at which the moon appears, and between two days this point may change so much on account of the moon's orbital motion as to make the retardation considerably greater or less than its average value. In northern latitudes this effect is particularly marked in the month of September, when the eastern horizon is nearly parallel with the moon's apparent path in the sky, and near the time of full moon in that month the moon rises on several successive nights at nearly the same hour, and in less degree the same is true for October. This highly convenient arrangement of moonlight has caused the full moons of these two months to be christened respectively the Harvest Moon and the Hunter's Moon.

94. ***Size and mass of the moon.***— It has been shown in Chapter I how the distance of the moon from the earth may be measured and its diameter determined by means of angles, and without enlarging upon the details of these observations, we note as their result that the moon is a globe 2,163 miles in diameter, and distant from the earth on the average about 240,000 miles. But, as we have seen in Chapter VII, this distance changes to the extent of a few thousand miles, sometimes less, sometimes greater, mainly on account of the elliptic shape of the moon's orbit about the earth, but also in part from the disturbing influence of other bodies, such as the sun, which pull the moon to and fro, backward and forward, to quite an appreciable extent.

From the known diameter of the moon it is a matter of elementary geometry to derive in miles the area of its surface and its volume or solid contents. Leaving this as an exercise for the student, we adopt the earth as the standard of comparison and find that the diameter of the moon is rather more than a quarter, 4/15, that of the earth, the area of its surface is a trifle more than 1/14 that of the earth, and its volume a little more than 1/49 of the earth's. So much is pure geometry, but we may combine with it some mechanical principles which enable us to go a step farther and to "weigh" the moon– i. e., determine its mass and the average density of the material of which it is made.

We have seen that the moon moves around the sun in a path differing but little from the smooth curve shown in Fig. 53, with arrows indicating the direction of motion, and it would follow absolutely such a smooth path were it not for the attraction of the earth, and in less degree of some of the other planets, which swing it about first to one side then to the other. But action and reaction are equal; the moon pulls as strongly upon the earth as does the earth upon the moon, and if earth and moon were of equal mass, the deviation of the earth from the smooth curve in the figure would be just as large as that of the moon. It is shown in the figure that the moon does displace the earth from this curve, and we have only to measure the amount of this displacement of the earth and compare it with the displacement suffered by the moon to find how much the mass of the one exceeds that of the other. It may be seen from the figure that at first quarter, about July 7th, the earth is thrust ahead in the direction of its orbital motion, while at the third quarter, July 22d, it is pulled back by the action of the moon, and at all times it is more or less displaced by this action, so that, in order to be strictly correct, we must amend our former statement about the moon moving around the earth and make it read, Both earth and moon revolve around a point on line between their centers. This point is called their *center of gravity*, and the earth and the moon both move in ellipses having this center of gravity at their common focus. Compare this with Kepler's First Law. These ellipses are similarly shaped, but of very different size, corresponding to Newton's third law of motion (Chapter IV), so that the action of the earth in causing the small moon to move around a large orbit is just equal to the reaction of the moon in causing the larger earth to move in the smaller orbit. This is equivalent to saying that the dimensions of the two orbits are inversely proportional to the masses of the earth and the moon.

By observing throughout the month the direction from the earth to the sun or to a near planet, such as Mars or Venus, astronomers have determined that the diameter of the ellipse in which the earth moves is about 5,850 miles, so that the distance of the earth from the center of gravity is 2,925 miles, and the distance of the moon from it is 240,000 - 2,925 = 237,075. We may now write in the form of a proportion–

Mass of earth : Mass of moon :: 237,075 : 2,925,

and find from it that the mass of the earth is 81 times as great as the mass of the moon– i. e., leaving kind and quality out of account, there is enough material in the earth to make 81 moons. We may note in this connection that the diameter of the earth, 7,926 miles, is greater than the diameter of the monthly orbit in which the moon causes it to move, and therefore the center of gravity of earth and moon always lies inside the body of the earth, about 1,000 miles below the surface.

95. ***Density of the moon.***— It is believed that in a general way the moon is made of much the same kind of material which goes to make up the earth— metals, minerals, rocks, etc.— and a part of the evidence upon which this belief is based lies in the density of the moon. By density of a substance we mean the amount of it which is contained in a given volume— i. e., the weight of a bushel or a cubic centimeter of the stuff. The density of chalk is twice as great as the density of water, because a cubic centimeter of chalk weighs twice as much as an equal volume of water, and similarly in other cases the density is found by dividing the mass or weight of the body by the mass or weight of an equal volume of water.

We know the mass of the earth (§ 45), and knowing the mass of a cubic foot of water, it is easy, although a trifle tedious, to compute what would be the mass of a volume of water equal in size to the earth. The quotient obtained by dividing one of these masses by the other (mass of earth ÷ mass of water) is the average density of the material composing the earth, and we find numerically that this is 5.6— i. e., it would take 5.6 water earths to attract as strongly as does the real one. From direct experiment we know that the average density of the principal rocks which make up the crust of the earth is only about half of this, showing that the deep-lying central parts of the earth are denser than the surface parts, as we should expect them to be, because they have to bear the weight of all that lies above them and are compressed by it.

Turning now to the moon, we find in the same way as for the earth that its average density is 3.4 as great as that of water.

96. ***Force of gravity upon the moon.***— This number, 3.4, compared with the 5.6 which we found for the earth, shows that on the whole the moon is made of lighter stuff than is the body of the earth, and this again is much what we should expect to find, for weight, the force which tends to compress the substance of the moon, is less there than here. The weight of a cubic yard of rock at the surface of either earth or moon is the force with which the earth or moon attracts it, and this by the law of gravitation is for the earth—

$W = k \cdot (m\, m')/(3963)^2;$

and for the moon—

$w = k \cdot \{m\,(m'/81)\}/(1081)^2;$

from which we find by division—

$w = (W/81)\,(3963/1081)^2 = W/6$ (approximately).

The cubic yard of rock, which upon the earth weighs two tons, would, if transported to the moon, weigh only one third of a ton, and would have only one sixth as much influence in compressing the rocks below it as it had upon the earth. Note that this rock when transported to the moon would be still attracted by the earth and would have weight toward the earth, but it is not this of which we are speaking; by its weight in the moon we mean the force with which the moon attracts it. Making due allowance for the difference in compression produced by weight, we may say that in general, so far as density goes, the moon is very like a piece of the earth of equal mass set off by itself alone.

97. ***Albedo.***– In another respect the lunar stuff is like that of which the earth is made: it reflects the sunlight in much the same way and to the same amount. The contrast of light and dark areas on the moon's surface shows, as we shall see in another section, the presence of different substances upon the moon which reflect the sunlight in different degrees. This capacity for reflecting a greater or less percentage of the incident sunlight is called *albedo* (Latin, whiteness), and the brilliancy of the full moon might lead one to suppose that its albedo is very great, like that of snow or those masses of summer cloud which we call thunderheads. But this is only an effect of contrast with the dark background of the sky. The same moon by day looks pale, and its albedo is, in fact, not very different from that of our common rocks– weather-beaten sandstone according to Sir John Herschel– so that it would be possible to build an artificial moon of rock or brick which would shine in the sunlight much as does the real moon.

The effect produced by the differences of albedo upon the moon's face is commonly called the "man in the moon," but, like the images presented by glowing coals, the face in the moon is anything which we choose to make it. Among the Chinese it is said to be a monkey pounding rice; in India, a rabbit; in Persia, the earth reflected as in a mirror, etc.

98. ***Librations.***– We have already learned that the moon turns always the same face toward the earth, and we have now to modify this statement and to find that here, as in so many other cases, the thing we learn first is only approximately true and needs to be limited or added to or modified in some way. In general, Nature is too complex to be completely understood at first sight or to be perfectly represented by a simple statement. In Fig. 55 we have two photographs of the moon, taken nearly three years apart, the right-hand one a little after first quarter and the left-hand one a little before third quarter. They therefore represent different parts of the moon's surface, but along the ragged edge the same region is shown on both photographs, and features common to both pictures may readily be found– e. g., the three rings which form a right-angled triangle about one third of the way down from the top of the cut, and the curved mountain chain just below these. If the moon turned exactly the same face toward us in the two pictures, the distance of any one of these markings from any part of the moon's edge must be the same in both pictures; but careful measurement will show that this is not the case, and that

in the left-hand picture the upper edge of the moon is tipped toward us and the lower edge away from us, as if the whole moon had been rotated slightly about a horizontal line and must be turned back a little (about 7°) in order to match perfectly the other part of the picture.

This turning is called a *libration*, and it should be borne in mind that the moon librates not only in the direction above measured, north and south, but also at right angles to this, east and west, so that we are able to see a little farther around every part of the moon's edge than would be possible if it turned toward us at all times exactly the same face. But in spite of the librations there remains on the farther side of the moon an area of 6,000,000 square miles which is forever hidden from us, and of whose character we have no direct knowledge, although there is no reason to suppose it very different from that which is visible, despite the fact that some of the books contain quaint speculations to the contrary. The continent of South America is just about equal in extent to this unknown region, while North America is a fair equivalent for all the rest of the moon's surface, both those central parts which are constantly visible, and the zone around the edge whose parts sometimes come into sight and are sometimes hidden.

An interesting consequence of the peculiar rotation of the moon is that from our side of it the earth is always visible. Sun, stars, and planets rise and set there as well as here, but to an observer on the moon the earth swings always overhead, shifting its position a few degrees one way or the other on account of the libration but running through its succession of phases, new earth, first quarter, etc., without ever going below the horizon, provided the observer is anywhere near the center of the moon's disk.

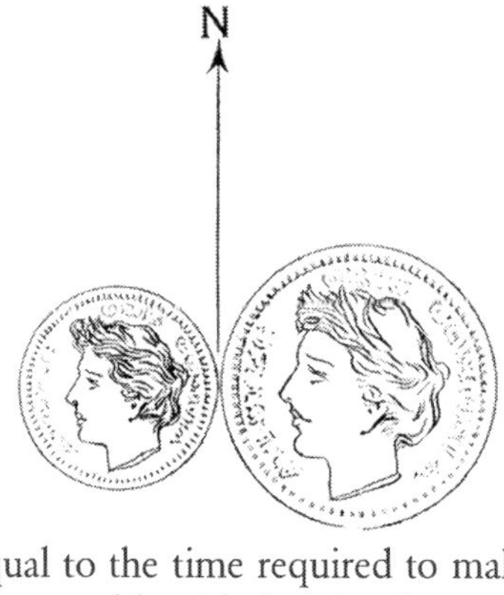

FIG. 54.—Illustrating the moon's rotation.

99. ***Cause of librations.***—That the moon should librate is by no means so remarkable a fact as that it should at all times turn very nearly the same face toward the earth. This latter fact can have but one meaning: the moon revolves about an axis as does the earth, but the time required for this revolution is just equal to the time required to make a revolution in its orbit. Place two coins upon a table with their heads turned toward the north, as in Fig. 54, and move the smaller one around the larger in such a way that its face shall always look away from the larger one. In making one revolution in its orbit the head on this small coin will be successively directed toward every point of the compass, and when it returns to its initial position the small coin will have

made just one revolution about an axis perpendicular to the plane of its orbit. In no other way can it be made to face always away from the figure at the center of its orbit while moving around it.

We are now in a position to understand the moon's librations, for, if the small coin at any time moves faster or slower in its orbit than it turns about its axis, a new side will be turned toward the center, and the same may happen if the central coin itself shifts into a new position. This is what happens to the moon, for its orbital motion, like that of Mercury (Fig. 17), is alternately fast and slow, and in addition to this there are present other minor influences, such as the fact that its rotation axis is not exactly perpendicular to the plane of its orbit; in addition to this the observer upon the earth is daily carried by its rotation from one point of view to another, etc., so that it is only in a general way that the rotation upon the axis and motion in the orbit keep pace with each other. In a general way a cable keeps a ship anchored in the same place, although wind and waves may cause it to "librate" about the anchor.

How the moon came to have this exact equality between its times of revolution and rotation constitutes a chapter of its history upon which we shall not now enter; but the equality having once been established, the mechanism by which it is preserved is simple enough.

The attraction of the earth for the moon has very slightly pulled the latter out of shape (§ 42), so that the particular diameter, which points toward the earth, is a little longer than any other, and thus serves as a handle which the earth lays hold of and pulls down into its lowest possible position– i. e., the position in which it points toward the center of the earth. Just how long this handle is, remains unknown, but it may be shown from the law of gravitation that less than a hundred yards of elongation would suffice for the work it has to do.

100. ***The moon as a world.***– Thus far we have considered the moon as a satellite of the earth, dependent upon the earth, and interesting chiefly because of its relation to it. But the moon is something more than this; it is a world in itself, very different from the earth, although not wholly unlike it. The most characteristic feature of the earth's surface is its division into land and water, and nothing of this kind can be found upon the moon. It is true that the first generation of astronomers who studied the moon with telescopes fancied that the large dark patches shown in Fig. 55 were bodies of water, and named them oceans, seas, lakes, and ponds, and to the present day we keep those names, although it is long since recognized that these parts of the moon's surface are as dry as any other. Their dark appearance indicates a different kind of material from that composing the lighter parts of the moon, material with a different albedo, just as upon the earth we have light-colored and dark-colored rocks, marble and slate, which seen from the moon must present similar contrasts of brightness. Although these dark patches are

almost the only features distinguishable with the unaided eye, it is far otherwise in the telescope or the photograph, especially along the ragged edge where great numbers of rings can be seen, which are apparently depressions in the moon and are called craters. These we find in great number all over the moon, but, as the figure shows, they are seen to the best advantage near the *terminator*— i. e., the dividing line between day and night, since the long shadows cast here by the rising or setting sun bring out the details of the surface better than elsewhere. Carefully examine Fig. 55 with reference to these features.

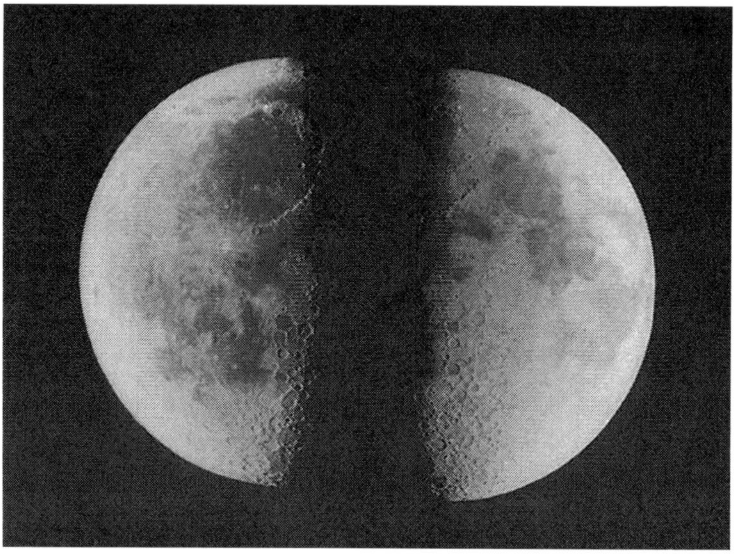

FIG. 55.— The moon at first and last quarter. Lick Observatory photographs.

Another feature which exists upon both earth and moon, although far less common there than here, is illustrated in the chain of mountains visible near the terminator, a little above the center of the moon in both parts of Fig. 55. This particular range of mountains, which is called the Lunar Apennines, is by far the most prominent one upon the moon, although others, the Alps and Caucasus, exist. But for the most part the lunar mountains stand alone, each by itself, instead of being grouped into ranges, as on the earth. Note in the figure that some of the lunar mountains stretch out into the night side of the moon, their peaks projecting up into the sunlight, and thus becoming visible, while the lowlands are buried in the shadow.

A subordinate feature of the moon's surface is the system of *rays* which seem to radiate like spokes from some of the larger craters, extending over hill and valley sometimes for hundreds of miles. A suggestion of these rays may be seen in Fig. 55, extending from the great crater Copernicus a little southwest of the end of the Apennines, but their most perfect development is to be seen at the time of full moon around the crater Tycho, which lies near the south pole of the moon. Look for them with an opera glass.

Another and even less conspicuous feature is furnished by the rills, which, under favorable conditions of illumination, appear like long cracks on the moon's surface, perhaps analogous to the cañons of our Western country.

101. ***The map of the moon.*** – Fig. 55 furnishes a fairly good map of a limited portion of the moon near the terminator, but at the edges little or no detail can be seen. This is always true; the whole of the moon can not be seen to advantage at any one time, and to remedy this we need to construct from many photographs or drawings a map which shall represent the several parts of the moon as they appear at their best. Fig. 56 shows such a map photographed from a relief model of the moon, and representing the principal features of the lunar surface in a way they can never be seen simultaneously. Perhaps its most striking feature is the shape of the craters, which are shown round in the central parts of the map and oval at the edges, with their long diameters parallel to the moon's edge. This is, of course, an effect of the curvature of the moon's surface, for we look very obliquely at the edge portions, and thus see their formations much foreshortened in the direction of the moon's radius.

FIG. 56.— Relief map of the moon's surface.— After NASMYTH and CARPENTER.

The north and south poles of the moon are at the top and bottom of the map respectively, and a mere inspection of the regions around them will show how much more rugged is the southern hemisphere of the moon than the northern. It furnishes, too, some indication of how numerous are the lunar craters, and how in crowded regions they overlap one another.

The student should pick out upon the map those features which he has learned to know in the photograph (Fig. 55)– the Apennines, Copernicus, and the continuation of the Apennines, extending into the dark part of the moon.

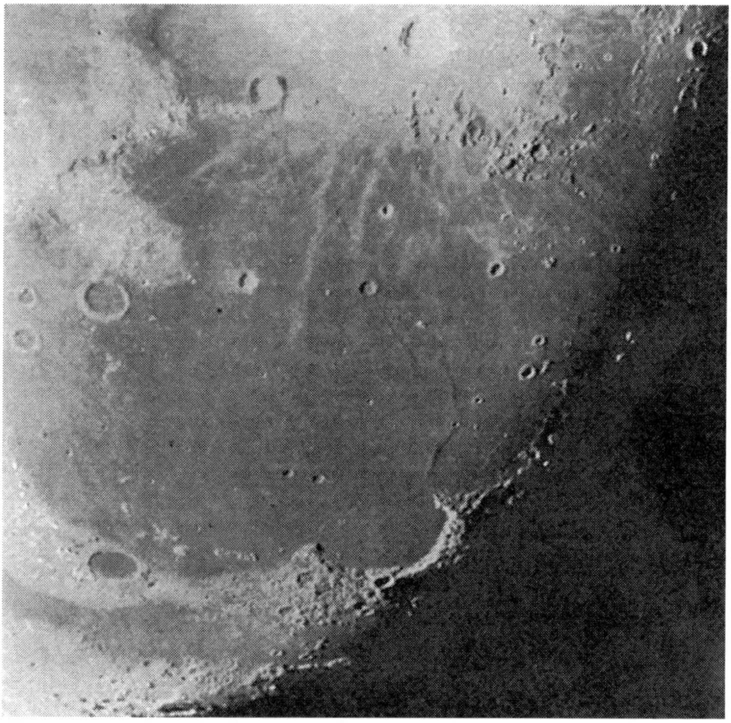

FIG. 57.— Mare Imbrium. Photographed by G. W. RITCHEY.

102. ***Size of the lunar features.*** — We may measure distances here in the same way as upon a terrestrial map, remembering that near the edges the scale of the map is very much distorted parallel to the moon's diameter, and measurements must not be taken in this direction, but may be taken parallel to the edge. Measuring with a millimeter scale, we find on the map for the diameter of the crater Copernicus, 2.1 millimeters. To turn this into the diameter of the real Copernicus in miles, we measure upon the same map the diameter of the moon, 79.7 millimeters, and then have the proportion—

Diameter of Copernicus in miles : 2,163 :: 2.1 : 79.7,

which when solved gives 57 miles. The real diameter of Copernicus is a trifle over 56 miles. At the eastern edge of the moon, opposite the Apennines, is a large oval spot called the Mare Crisium (Latin, *ma-re* = sea). Measure its length. The large crater to the northwest of the Apennines is called Archimedes. Measure its diameter both in the map and in the photograph (Fig. 55), and see how the two results agree. The true diameter of this crater, east and west, is very approximately 50 miles. The great smooth surface to the west of Archimedes is the Mare Imbrium.

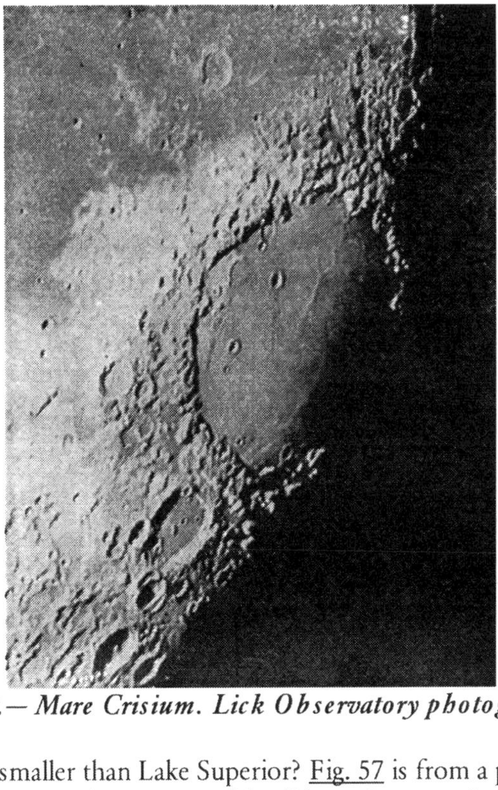

FIG. 58.— Mare Crisium. Lick Observatory photographs.

Is it larger or smaller than Lake Superior? Fig. 57 is from a photograph of the Mare Imbrium, and the amount of detail here shown at the bottom of the sea is a sufficient indication that, in this case at least, the water has been drawn off, if indeed any was ever present.

Fig. 58 is a representation of the Mare Crisium at a time when night was beginning to encroach upon its eastern border, and it serves well to show the rugged character of the ring-shaped wall which incloses this area.

With these pictures of the smoother parts of the moon's surface we may compare Fig. 59, which shows a region near the north pole of the moon, and Fig. 60, giving an early morning view of Archimedes and the Apennines. Note how long and sharp are the shadows.

*FIG. 59.— **Illustrating the rugged character of the moon's surface.**— NASMYTH and CARPENTER.*

103. ***The moon's atmosphere.***— Upon the earth the sun casts no shadows so sharp and black as those of Fig. 60, because his rays are here scattered and reflected in all directions by the dust and vapors of the atmosphere (§ 51), so that the place from which direct sunlight is cut off is at least partially illumined by this reflected light. The shadows of Fig. 60 show that upon the moon it must be otherwise, and suggest that if the moon has any atmosphere whatever, its density must be utterly insignificant in comparison with that of the earth. In its motion around the earth the moon frequently eclipses stars (*occults* is the technical word), and if the moon had an atmosphere such as is shown in Fig. 61, the light from the star *A* must shine through this atmosphere just before the moon's advancing body cuts it off, and it must be refracted by the atmosphere so that the star would appear in a slightly different direction (nearer to *B*) than before. The earth's atmosphere refracts the starlight under such circumstances by more than a degree, but no one has been able to find in the case of the moon any effect of this kind amounting to even a fraction of a second of arc. While this hardly justifies the statement sometimes made that the moon has no atmosphere, we shall be entirely safe in saying that if it has one at all its density is less than a

thousandth part of that of the earth's atmosphere. Quite in keeping with this absence of an atmosphere is the fact that clouds never float over the surface of the moon. Its features always stand out hard and clear, without any of that haze and softness of outline which our atmosphere introduces into all terrestrial landscapes.

FIG. 60.— Archimedes and Apennines. NASMYTH and CARPENTER.

104. ***Height of the lunar mountains.***— Attention has already been called to the detached mountain peaks, which in Fig. 55 prolong the range of Apennines into the lunar night. These are the beginnings of the Caucasus mountains, and from the photograph we may measure as follows the height to which they rise above the surrounding level of the moon: Fig. 62 represents a part of the lunar surface along the boundary line between night and day, the horizontal line at the top of the figure representing a level ray of sunlight which just touches the moon at T and barely illuminates the top of the mountain, M, whose height, h, is to be determined. If we let R stand for the radius of the moon and s for the distance, TM, we shall have in the right-angled triangle MTC,

$$R^2 + s^2 = (R + h)^2,$$

and we need only to measure s— that is, the distance from the terminator to the detached mountain peak— to make this equation determine h, since R is already known, being half the diameter of the moon— 1,081 miles. Practically it is more convenient to use instead of this equation another form, which the student who is expert in algebra may show to be very nearly equivalent to it:

$$h\,(\text{miles}) = s^2 / 2163,$$

$$\text{or } h\,(\text{feet}) = 2.44\, s^2.$$

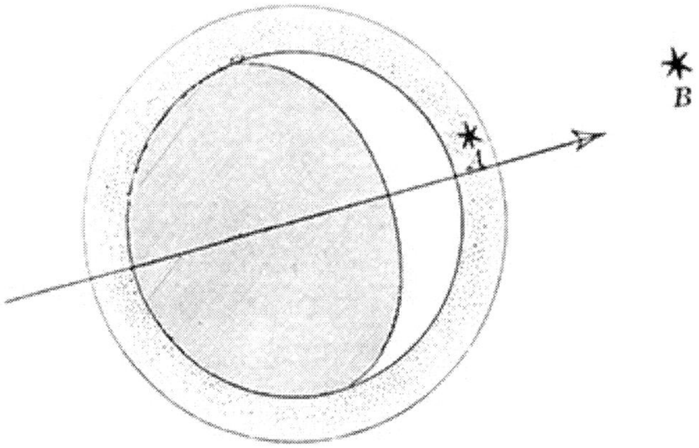

FIG. 61.—Occultations and the moon's atmosphere.

The distance *s* must be expressed in miles in all of these equations. In Fig. 55 the distance from the terminator to the first detached peak of the Caucasus mountains is 1.7 millimeters = 52 miles, from which we find the height of the mountain to be 1.25 miles, or 6,600 feet.

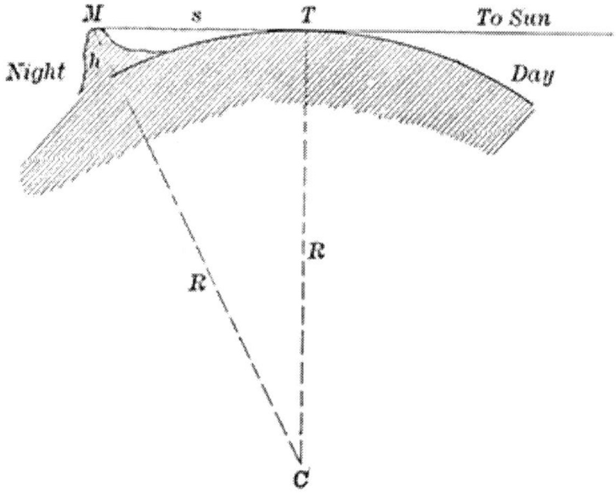

FIG. 62.—Determining the height of a lunar mountain.

Two things, however, need to be borne in mind in this connection. On the earth we measure the heights of mountains *above sea level*, while on the moon there is no sea, and our 6,600 feet is simply the height of the mountain top above the level of that particular point in the terminator, from which we measure its distance. So too it is evident from the appearance of things, that the sunlight, instead of just touching the top of the particular mountain whose height we have measured, really extends some little distance down from its summit, and the 6,600 feet is therefore the elevation of the lowest point on the mountains to which the sunlight reaches. The peak itself may be several hundred feet higher, and our photograph must be taken at the exact moment when this peak appears in the lunar morning or disappears in the evening if we are to measure the altitude of the mountain's summit. Measure the height of the most northern visible mountain of the Caucasus range. This is one of the outlying spurs of the great mountain Calippus, whose principal peak, 19,000 feet high, is shown in Fig. 55 as the brightest part of the Caucasus range.

The highest peak of the lunar Apennines, Huyghens, has an altitude of 18,000 feet, and the Leibnitz and Doerfel Mountains, near the south pole of the moon, reach an altitude 50 per cent greater than this, and are probably the highest peaks on the moon. This falls very little short of the highest mountain on the earth, although the moon is much smaller than the earth, and these mountains are considerably higher than anything on the western continent of the earth.

The vagueness of outline of the terminator makes it difficult to measure from it with precision, and somewhat more accurate determinations of the heights of lunar mountains can be obtained by measuring the length of the shadows which they cast, and the depths of craters may also be measured by means of the shadows which fall into them.

105. ***Craters.***— Fig. 63 shows a typical lunar crater, and conveys a good idea of the ruggedness of the lunar landscape. Compare the appearance of this crater with the following generalizations, which are based upon the accurate measurement of many such:

A. A crater is a real depression in the surface of the moon, surrounded usually by an elevated ring which rises above the general level of the region outside, while the bottom of the crater is about an equal distance below that level.

B. Craters are shallow, their diameters ranging from five times to more than fifty times their depth. Archimedes, whose diameter we found to be 50 miles, has an average depth of about 4,000 feet below the crest of its surrounding wall, and is relatively a shallow crater.

C. Craters frequently have one or more hills rising within them which, however, rarely, if ever, reach up to the level of the surrounding wall.

D. Whatever may have been the mode of their formation, the craters can not have been produced by scooping out material from the center and piling it up to make the wall, for in three cases out of four the volume of the excavation is greater than the volume of material contained in the wall.

FIG. 63.—A typical lunar crater.—NASMYTH and CARPENTER.

106. ***Moon and earth.***— We have gone far enough now to appreciate both the likeness and the unlikeness of the moon and earth. They may fairly enough be likened to offspring of the same parent who have followed very different careers, and in the fullness of time find themselves in very different circumstances. The most serious point of difference in these circumstances is the atmosphere, which gives to the earth a wealth of phenomena altogether lacking in the moon. Clouds, wind, rain, snow, dew, frost, and hail are all dependent upon the atmosphere and can not be found where it is not. There can be nothing upon the moon at all like that great group of changes which we call weather, and the unruffled aspect of the moon's face contrasts sharply with the succession of cloud and sunshine which the earth would present if seen from the moon.

The atmosphere is the chief agent in the propagation of sound, and without it the moon must be wrapped in silence more absolute than can be found upon the surface of the earth. So, too, the absence of an atmosphere shows that there can be no water or other liquid upon the moon, for if so it would immediately evaporate and produce a gaseous envelope which we have seen does not exist. With air and water absent there can be of course no vegetation or life of any kind upon the moon, and we are compelled to regard it as an arid desert, utterly waste.

107. ***Temperature of the moon.***— A characteristic feature of terrestrial deserts, which is possessed in exaggerated degree by the moon, is the great extremes of temperature to which they and it are subject. Owing to its slow rotation about its axis, a point on the moon receives the solar radiation uninterruptedly for more than a fortnight, and that too unmitigated by any cloud or vaporous covering. Then for a like period it is turned away from the sun and allowed to cool off, radiating into interplanetary space without hindrance its accumulated store of heat. It is easy to see that the range of temperature between day and night must be much greater under these circumstances than it is with us where shorter days and clouded skies render day and night more nearly alike, to say nothing of the ocean whose waters serve as a great balance wheel for equalizing temperatures. Just how hot or how cold the moon becomes is hard to determine, and very different estimates are to be found in the books. Perhaps the most reliable of these are furnished by the recent researches of Professor Very, whose experiments lead him to conclude that "its rocky surface at midday, in latitudes where the sun is high, is probably hotter than boiling water and only the most terrible of earth's deserts, where the burning sands blister the skin, and men, beasts, and birds drop dead, can approach a noontide on the cloudless surface of our satellite. Only the extreme polar latitudes of the moon can have an endurable temperature by day, to say nothing of the night, when we should have to become troglodytes to preserve ourselves from such intense cold."

While the night temperature of the moon, even very soon after sunset, sinks to something like 200° below zero on the centigrade scale, or 320° below zero on the Fahrenheit scale, the lowest known temperature upon the earth, according to General Greely, is 90° Fahr. below zero, recorded in Siberia in January, 1885.

Winter and summer are not markedly different upon the moon, since its rotation axis is nearly perpendicular to the plane of the earth's orbit about the sun, and the sun never goes far north or south of the moon's equator. The month is the one cycle within which all seasonal changes in its physical condition appear to run their complete course.

108. ***Changes in the moon.***— It is evidently idle to look for any such changes in the condition of the moon's surface as with us mark the progress of the seasons or the spread of civilization over the wilderness. But minor changes there may be, and it would seem that the violent oscillations of temperature from day to night ought to have some effect in breaking down and crumbling the sharp peaks and crags which are there so common and so pronounced. For a century past astronomers have searched carefully for changes of this kind— the filling up of some crater or the fall of a mountain peak; but while some things of this kind have been reported from time to time, the evidence in their behalf has not been altogether conclusive. At the present time it is an open question whether changes of this sort large enough to be seen from the earth are in progress. A crater much less than a mile wide can be seen in the telescope, but it is not easy to tell whether so minute an object has changed in size or shape during a year or a decade, and even if changes are seen they may be apparent rather than

real. Fig. 64 contains two views of the crater Archimedes, taken under a morning and an afternoon sun respectively, and shows a very pronounced difference between the two which proceeds solely from a difference of illumination. In the presence of such large fictitious changes astronomers are slow to accept smaller ones as real.

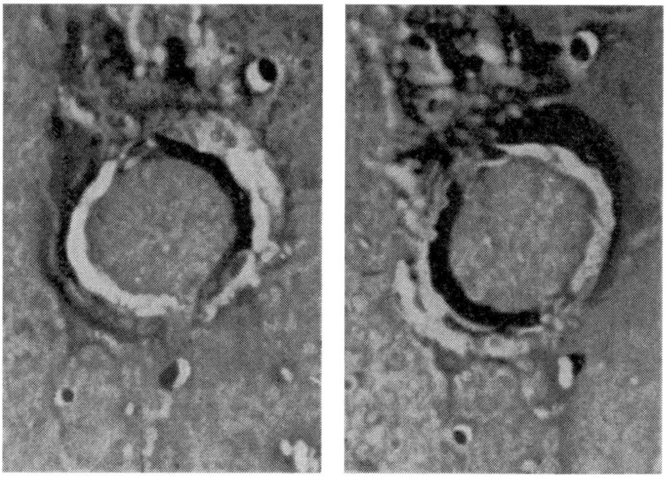

FIG. 64.—Archimedes in the lunar morning and afternoon.—WEINEK.

It is this absence of change that is responsible for the rugged and sharp-cut features of the moon which continue substantially as they were made, while upon the earth rain and frost are continually wearing down the mountains and spreading their substance upon the lowland in an unending process of smoothing off the roughnesses of its surface. Upon the moon this process is almost if not wholly wanting, and the moon abides to-day much more like its primitive condition than is the earth.

109. ***The moon's influence upon the earth.***— There is a widespread popular belief that in many ways the moon exercises a considerable influence upon terrestrial affairs: that it affects the weather for good or ill, that crops must be planted and harvested, pigs must be killed, and timber cut at the right time of the moon, etc. Our common word lunatic means moonstruck— i. e., one upon whom the moon has shone while sleeping. There is not the slightest scientific basis for any of these beliefs, and astronomers everywhere class them with tales of witchcraft, magic, and popular delusion. For the most part the moon's influence upon the earth is limited to the light which it sends and the effect of its gravitation, chiefly exhibited in the ocean tides. We receive from the moon a very small amount of second-hand solar heat and there is also a trifling magnetic influence, but neither of these last effects comes within the range of ordinary observation, and we shall not go far wrong in

saying that, save the moonlight and the tides, every supposed lunar influence upon the earth is either fictitious or too small to be readily detected.

CHAPTER X.
THE SUN

110. ***Dependence of the earth upon the sun.*** – There is no better introduction to the study of the sun than Byron's Ode to Darkness, beginning with the lines–

"I dreamed a dreamThat was not all a dream.The bright sun was extinguished,"

and proceeding to depict in vivid words the consequences of this extinction. The most matter-of-fact language of science agrees with the words of the poet in declaring the earth's dependence upon the sun for all those varied forms of energy which make it a fit abode for living beings. The winds blow and the rivers run; the crops grow, are gathered and consumed, by virtue of the solar energy. Factory, locomotive, beast, bird, and the human body furnish types of machines run by energy derived from the sun; and the student will find it an instructive exercise to search for kinds of terrestrial energy which are not derived either directly or indirectly from the sun. There are a few such, but they are neither numerous nor important.

111. ***The sun's distance from the earth.*** – To the astronomer the sun presents problems of the highest consequence and apparently of very diverse character, but all tending toward the same goal: the framing of a mechanical explanation of the sun considered as a machine; what it is, and how it does its work. In the forefront of these problems stand those numerical determinations of distance, size, mass, density, etc., which we have already encountered in connection with the moon, but which must here be dealt with in a different manner, because the immensely greater distance of the sun makes impossible the resort to any such simple method as the triangle used for determining the moon's distance. It would be like determining the distance of a steeple a mile away by observing its direction first from one eye, then from the other; too short a base for the triangle. In one respect, however, we stand upon a better footing than in the case of the moon, for the mass of the earth has already been found (Chapter IV) as a fractional part of the sun's mass, and we have only to invert the fraction in order to find that the sun's mass is 329,000 times that of the earth and moon combined, or 333,000 times that of the earth alone.

If we could rely implicitly upon this number we might make it determine for us the distance of the sun through the law of gravitation as follows: It was suggested in § 38 that Newton proved Kepler's three laws to be imperfect corollaries from the law of gravitation, requiring a little amendment to make them strictly correct, and below we give in the form of an equation Kepler's statement of the Third Law together with Newton's amendment of it. In these equations–

T = Periodic time of any planet;

a = One half the major axis of its orbit;
m = Its mass;
M = The mass of the sun;
k = The gravitation constant corresponding to the particular set of units in which T, a, m, and M are expressed.

(Kepler) $a^3/T^2 = h$; (Newton) $a^3/T^2 = k\,(M + m)$.

Kepler's idea was: For every planet which moves around the sun, a^3 divided by T^2 always gives the same quotient, h; and he did not concern himself with the significance of this quotient further than to note that if the particular a and T which belong to any planet– e. g., the earth– be taken as the units of length and time, then the quotient will be 1. Newton, on the other hand, attached a meaning to the quotient, and showed that it is equal to the product obtained by multiplying the sum of the two masses, planet and sun, by a number which is always the same when we are dealing with the action of gravitation, whether it be between the sun and planet, or between moon and earth, or between the earth and a roast of beef in the butcher's scales, provided only that we use always the same units with which to measure times, distances, and masses.

Numerically, Newton's correction to Kepler's Third Law does not amount to much in the motion of the planets. Jupiter, which shows the greatest effect, makes the circuit of his orbit in 4,333 days instead of 4,335, which it would require if Kepler's law were strictly true. But in another respect the change is of the utmost importance, since it enables us to extend Kepler's law, which relates solely to the sun and its planets, to other attracting bodies, such as the earth, moon, and stars. Thus for the moon's motion around the earth we write–

$(240{,}000)^3/(27.32)^2 = k\,(1 + 1/81)$,

from which we may find that, with the units here employed, the earth's mass as the unit of mass, the mean solar day as the unit of time, and the mile as the unit of distance–

$k = 1830 \times 10^{10}$.

If we introduce this value of k into the corresponding equation, which represents the motion of the earth around the sun, we shall have–

$a^3/(365.25)^2 = 1830 \times 10^{10}\,(333{,}000 + 1)$,

where the large number in the parenthesis represents the number of times the mass of the sun is greater than the mass of the earth. We shall find by solving this equation that a, the mean distance of the sun from the earth, is very approximately 93,000,000 miles.

113. ***Another method of determining the sun's distance.***– This will be best appreciated by a reference to Fig. 17. It appears here that the earth makes its nearest approach to the orbit of Mars in the month of August, and if in any August Mars happens to be in opposition, its distance from the earth will

be very much less than the distance of the sun from the earth, and may be measured by methods not unlike those which served for the moon. If now the orbits of Mars and the earth were circles having their centers at the sun this distance between them, which we may represent by D, would be the difference of the radii of these orbits—

$$D = a'' - a',$$

where the accents ", ' represent Mars and the earth respectively. Kepler's Third Law furnishes the relation—

$$(a'')^3/(T'')^2 = (a')^3/(T')^2;$$

and since the periodic times of the earth and Mars, T', T'', are known to a high degree of accuracy, these two equations are sufficient to determine the two unknown quantities, a', a''— i. e., the distance of the sun from Mars as well as from the earth. The first of these equations is, of course, not strictly true, on account of the elliptical shape of the orbits, but this can be allowed for easily enough.

In practice it is found better to apply this method of determining the sun's distance through observations of an asteroid rather than observations of Mars, and great interest has been aroused among astronomers by the discovery, in 1898, of an asteroid, or planet, Eros, which at times comes much closer to the earth than does Mars or any other heavenly body except the moon, and which will at future oppositions furnish a more accurate determination of the sun's distance than any hitherto available. Observations for this purpose are being made at the present time (October, 1900).

Many other methods of measuring the sun's distance have been devised by astronomers, some of them extremely ingenious and interesting, but every one of them has its weak point— e. g., the determination of the mass of the earth in the first method given above and the measurement of D in the second method, so that even the best results at present are uncertain to the extent of 200,000 miles or more, and astronomers, instead of relying upon any one method, must use all of them, and take an average of their results. According to Professor Harkness, this average value is 92,796,950 miles, and it seems certain that a line of this length drawn from the earth toward the sun would end somewhere within the body of the sun, but whether on the nearer or the farther side of the center, or exactly at it, no man knows.

114. ***Parallax and distance.***— It is quite customary among astronomers to speak of the sun's parallax, instead of its distance from the earth, meaning by parallax its difference of direction as seen from the center and surface of the earth— i. e., the angle subtended at the sun by a radius of the earth placed at right angles to the line of sight. The greater the sun's distance the smaller will this angle be, and it therefore makes a substitute for the distance which has the advantage of being represented by a small number, 8".8, instead of a large one.

The books abound with illustrations intended to help the reader comprehend how great is a distance of 93,000,000 miles, but a single one of these must suffice here. To ride 100 miles a day 365 days in the year would be counted a good bicycling record, but the rider who started at the beginning of the Christian era and rode at that rate toward the sun from the year 1 A. D. down to the present moment would not yet have reached his destination, although his journey would be about three quarters done. He would have crossed the orbit of Venus about the time of Charlemagne, and that of Mercury soon after the discovery of America.

115. ***Size and density of the sun.*** – Knowing the distance of the sun, it is easy to find from the angle subtended by its diameter (32 minutes of arc) that the length of that diameter is 865,000 miles. We recall in this connection that the diameter of the moon's *orbit* is only 480,000 miles, but little more than half the diameter of the sun, thus affording abundant room inside the sun, and to spare, for the moon to perform the monthly revolution about its orbit, as shown in Fig. 65.

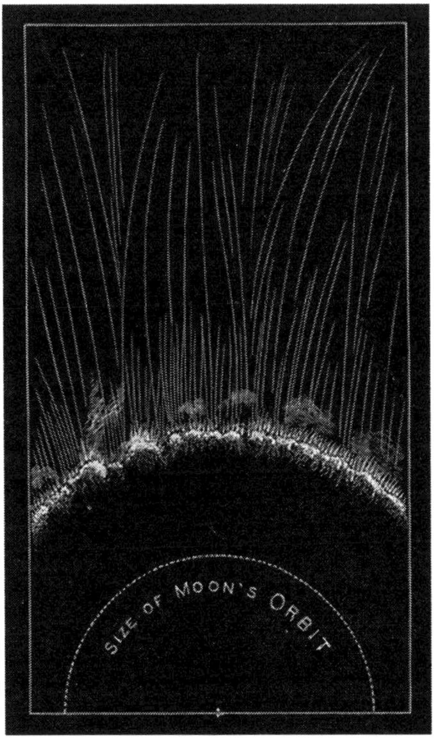

FIG. 65.— The sun's size.— YOUNG.

In the same manner in which the density of the moon was found from its mass and diameter, the student may find from the mass and diameter of the sun given above that its mean density is 1.4 times that of water. This is about the same as the density of gravel or soft coal, and is just about one quarter of the average density of the earth.

We recall that the small density of the moon was accounted for by the diminished weight of objects upon it, but this explanation can not hold in the case of the sun, for not only is the density less but the force of gravity (weight) is there 28 times as great as upon the earth. The athlete who here weighs 175 pounds, if transported to the surface of the sun would weigh more than an elephant does here, and would find his bones break under his own weight if his muscles were strong enough to hold him upright. The tremendous pressure exerted by gravity at the surface of the sun must be surpassed below the surface, and as it does not pack the material together and make it dense, we are driven to one of two conclusions: Either the stuff of which the sun is made is altogether unlike that of the earth, not so readily compressed by pressure, or there is some opposing influence at work which more than balances the effect of gravity and makes the solar stuff much lighter than the terrestrial.

116. ***Material of which the sun is made.***— As to the first of these alternatives, the spectroscope comes to our aid and shows in the sun's spectrum (Fig. 50) the characteristic line marked *D*, which we know always indicates the presence of sodium and identifies at least one terrestrial substance as present in the sun in considerable quantity. The lines marked *C* and *F* are produced by hydrogen, which is one of the constituents of water, *E* shows calcium to be present in the sun, *b* magnesium, etc. In this way it has been shown that about one half of our terrestrial elements, mainly the metallic ones, are present as gases on or near the sun's surface, but it must not be inferred that elements not found in this way are absent from the sun. They may be there, probably are there, but the spectroscopic proof of their presence is more difficult to obtain. Professor Rowland, who has been prominent in the study of the solar spectrum, says: "Were the whole earth heated to the temperature of the sun, its spectrum would probably resemble that of the sun very closely."

Some of the common terrestrial elements found in the sun are:

Aluminium.
Calcium.
Carbon.
Copper.
Hydrogen.
Iron.
Lead.
Nickel.
Potassium.
Silicon.
Silver.
Sodium.
Tin.
Zinc.
Oxygen (?)

Whatever differences of chemical structure may exist between the sun and the earth, it seems that we must regard these bodies as more like than unlike to each other in substance, and we are brought back to the second of our alternatives: there must be some influence opposing the force of gravity and making the substance of the sun light instead of heavy, and we need not seek far to find it in—

117. ***The heat of the sun.***— That the sun is hot is too evident to require proof, and it is a familiar fact that heat expands most substances and makes them less dense. The sun's heat falling upon the earth expands it and diminishes its density in some small degree, and we have only to imagine this process of expansion continued until the earth's diameter becomes 58 per cent larger than it now is, to find the earth's density reduced to a level with that of the sun. Just how much the temperature of the earth must be raised to produce this amount of expansion we do not know, neither do we know accurately the temperature of the sun, but there can be no doubt that heat is the cause of the sun's low density and that the corresponding temperature is very high.

Before we inquire more closely into the sun's temperature, it will be well to draw a sharp distinction between the two terms heat and temperature, which are often used as if they meant the same thing. Heat is a form of energy which may be found in varying degree in every substance, whether warm or cold— a block of ice contains a considerable amount of heat— while temperature corresponds to our sensations of warm and cold, and measures the extent to which heat is concentrated in the body. It is the amount of heat per molecule of the body. A barrel of warm water contains more heat than the flame of a match, but its temperature is not so high. Bearing in mind this distinction, we seek to determine not the amount of heat contained in the sun but the sun's temperature, and this involves the same difficulty as does the question, What is the temperature of a locomotive? It is one thing in the fire box and another thing in the driving wheels, and still another at the headlight; and so with the sun, its temperature is certainly different in different parts— one thing at the center and another at the surface. Even those parts which we see are covered by a veil of gases which produce by absorption the dark lines of the solar spectrum, and seriously interfere both with the emission of energy from the sun and with our attempts at measuring the temperature of those parts of the surface from which that energy streams.

In view of these and other difficulties we need not be surprised that the wildest discordance has been found in estimates of the solar temperature made by different investigators, who have assigned to it values ranging from 1,400° C. to more than 5,000,000° C. Quite recently, however, improved methods and a better understanding of the problem have brought about a better agreement of results, and it now seems probable that the temperature of the visible surface of the sun lies somewhere between 5,000° and 10,000° C., say 15,000° of the Fahrenheit scale.

118. ***Determining the sun's temperature.***– One ingenious method which has been used for determining this temperature is based upon the principle stated above, that every object, whether warm or cold, contains heat and gives it off in the form of radiant energy. The radiation from a body whose temperature is lower than 500° C. is made up exclusively of energy whose wave length is greater than 7,600 tenth meters, and is therefore invisible to the eye, although a thermometer or even the human hand can often detect it as radiant heat. A brick wall in the summer sunshine gives off energy which can be felt as heat but can not be seen. When such a body is further heated it continues to send off the same kinds (wave lengths) of energy as before, but new and shorter waves are added to its radiation, and when it begins to emit energy of wave length 7,500 or 7,600 tenth meters, it also begins to shine with a dull-red light, which presently becomes brighter and less ruddy and changes to white as the temperature rises, and waves of still shorter length are thereby added to the radiation. We say, in common speech, the body becomes first red hot and then white hot, and we thus recognize in a general way that the kind or color of the radiation which a body gives off is an index to its temperature. The greater the proportion of energy of short wave lengths the higher is the temperature of the radiating body. In sunlight the maximum of brilliancy to the eye lies at or near the wave length, 5,600 tenth meters, but the greatest intensity of radiation of all kinds (light included) is estimated to fall somewhere between green and blue in the spectrum at or near the wave length 5,000 tenth meters, and if we can apply to this wave length Paschen's law– temperature reckoned in degrees centigrade from the absolute zero is always equal to the quotient obtained by dividing the number 27,000,000 by the wave length corresponding to maximum radiation– we shall find at once for the absolute temperature of the sun's surface 5,400° C.

Paschen's law has been shown to hold true, at least approximately, for lower temperatures and longer wave lengths than are here involved, but as it is not yet certain that it is strictly true and holds for all temperatures, too great reliance must not be attached to the numerical result furnished by it.

119. ***The sun's surface.***– A marked contrast exists between the faces of sun and moon in respect of the amount of detail to be seen upon them, the sun showing nothing whatever to correspond with the mountains, craters, and seas of the moon. The unaided eye in general finds in the sun only a blank bright circle as smooth and unmarked as the surface of still water, and even the telescope at first sight seems to show but little more.

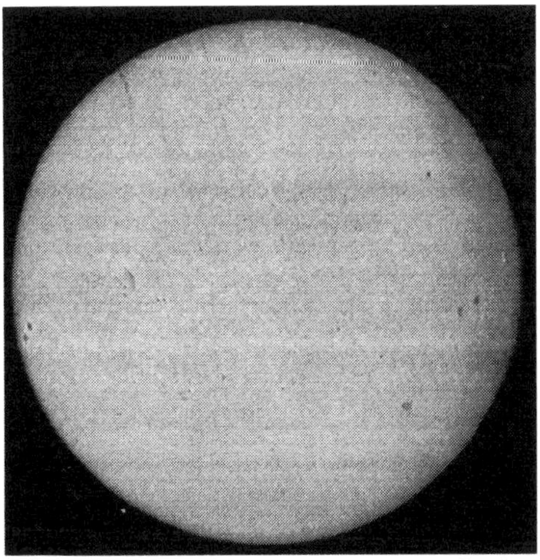

FIG. 66.— The sun, August 11, 1894. Photographed at the Goodsell Observatory.

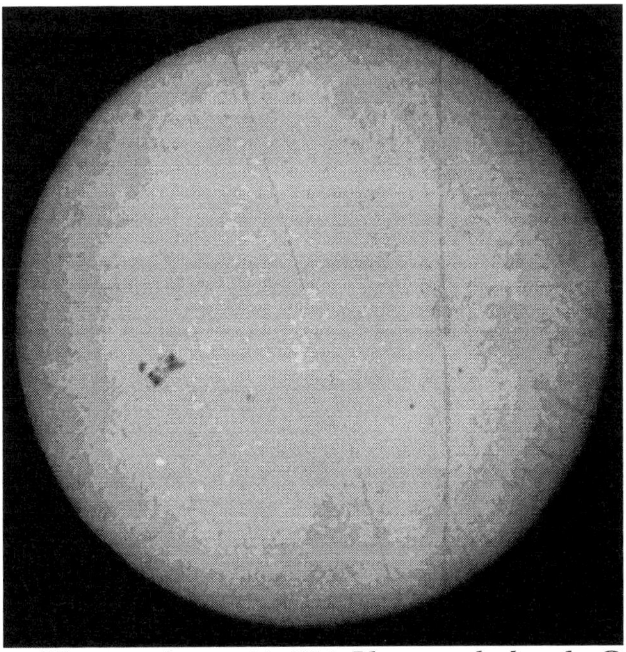

FIG. 67.— The sun, August 14, 1894. Photographed at the Goodsell Observatory.

There may usually be found upon the sun's face a certain number of black patches called *sun spots*, such as are shown in Figs. 66 to 69, and occasionally these are large enough to be seen through a smoked glass without the aid of a telescope. When seen near the edge of the sun they are quite frequently accompanied, as in Fig. 69, by vague patches called *faculæ* (Latin, *facula* = a little torch), which look a little brighter than the surrounding parts of the sun. So, too, a good photograph of the sun usually shows that the central parts of the disk are rather brighter than the edge, as indeed we should expect them to be, since the absorption lines in the sun's spectrum have already taught us that the visible surface of the sun is enveloped by invisible vapors which in some measure absorb the emitted light and render it feebler at the edge where it passes through a greater thickness of this envelope than at the center. See Fig. 70, where it is shown that the energy coming from the edge of the sun to the earth has to traverse a much longer path inside the vapors than does that coming from the center.

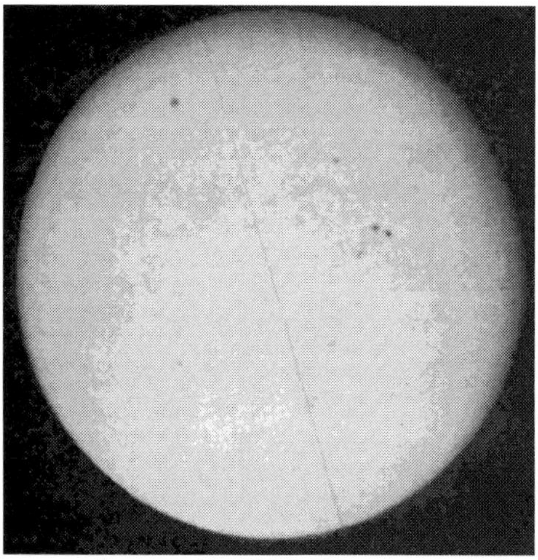

FIG. 68.— The sun, August 18, 1894. Photographed at the Goodsell Observatory.

Examine the sun spots in the four photographs, Figs. 66 to 69, and note that the two spots which appear at the extreme left of the first photograph, very much distorted and foreshortened by the curvature of the sun's surface, are seen in a different part of the second picture, and are not only more conspicuous but show better their true shape.

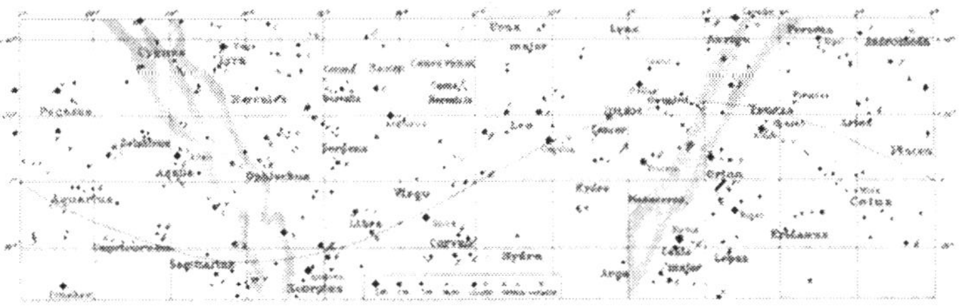

THE EQUATORIAL CONSTELLATIONS

Plate Ii. The Equatorial Constellations

120. ***The sun's rotation.***– The changed position of these spots shows that the sun rotates about an axis at right angles to the direction of the spot's motion, and the position of this axis is shown in the figure by a faint line ruled obliquely across the face of the sun nearly north and south in each of the four photographs. This rotation in the space of three days has carried the spots from the edge halfway to the center of the disk, and the student should note the progress of the spots in the two later photographs, that of August 21st showing them just ready to disappear around the farther edge of the sun.

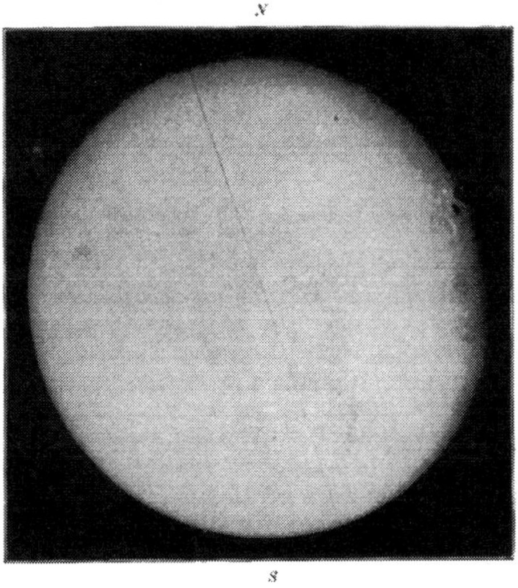

FIG. 69.— The sun, August 21, 1894. Photographed at the Goodsell Observatory.

Plot accurately in one of these figures the positions of the spots as shown in the other three, and observe whether the path of the spots across the sun's face is a straight line. Is there any reason why it should not be straight?

These four pictures may be made to illustrate many things about the sun. Thus the sun's axis is not parallel to that of the earth, for the letters *NS* mark the direction of a north and south line across the face of the sun, and this line, of course, is parallel to the earth's axis, while it is evidently not parallel to the sun's axis. The group of spots took more than ten days to move across the sun's face, and as at least an equal time must be required to move around the opposite side of the sun, it is evident that the period of the sun's rotation is something more than 20 days. It is, in fact, rather more than 25 days, for this same group of spots reappeared again on the left-hand edge of the sun on September 5th.

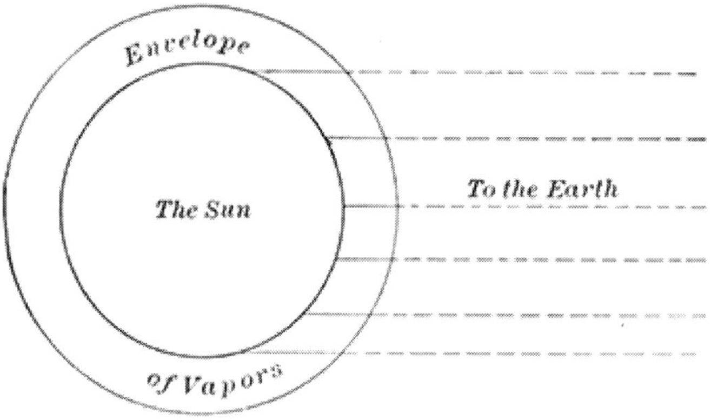

FIG. 70.— Absorption at the sun's edge.

121. ***Sun spots.***— Another significant fact comes out plainly from the photographs. The spots are not permanent features of the sun's face, since they changed their size and shape very appreciably in the few days covered by the pictures. Compare particularly the photographs of August 14th and August 18th, where the spots are least distorted by the curvature of the sun's surface. By September 16th this group of spots had disappeared absolutely from the sun's face, although when at its largest the group extended more than 80,000 miles in length, and several of the individual spots were large enough to contain the earth if it had been dropped upon them. From Fig. 67 determine in miles the length of the group on August 14th. Fig. 71 shows an enlarged view of these spots as they appeared on August 17th, and in this we find some details not so well shown in the preceding pictures. The larger spots consist of a black part called the *nucleus* or *umbra* (Latin, shadow), which is surrounded by an irregular

border called the *penumbra* (partial shadow), which is intermediate in brightness between the nucleus and the surrounding parts of the sun. It should not be inferred from the picture that the nucleus is really black or even dark. It shines, in fact, with a brilliancy greater than that of an electric lamp, but the background furnished by the sun's surface is so much brighter that by contrast with it the nucleus and penumbra appear relatively dark.

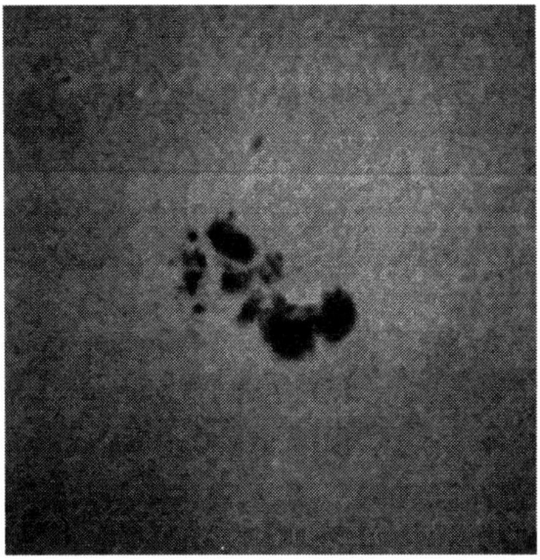

FIG. 71.— Sun spots, August 17, 1894. Goodsell Observatory.

FIG. 72.— Sun spot of March 5, 1873.— From LANGLEY, The New Astronomy. By permission of the publishers.

The bright shining surface of the sun, the background for the spots, is called the *photosphere* (Greek, light sphere), and, as Fig. 71 shows, it assumes under a suitable magnifying power a mottled aspect quite different from the featureless expanse shown in the earlier pictures. The photosphere is, in fact, a layer of little clouds with darker spaces between them, and the fine detail of these clouds, their complicated structure, and the way in which, when projected against the background of a sun spot, they produce its penumbra, are all brought out in Fig. 72. Note that the little patch in one corner of this picture represents North and South America drawn to the same scale as the sun spots.

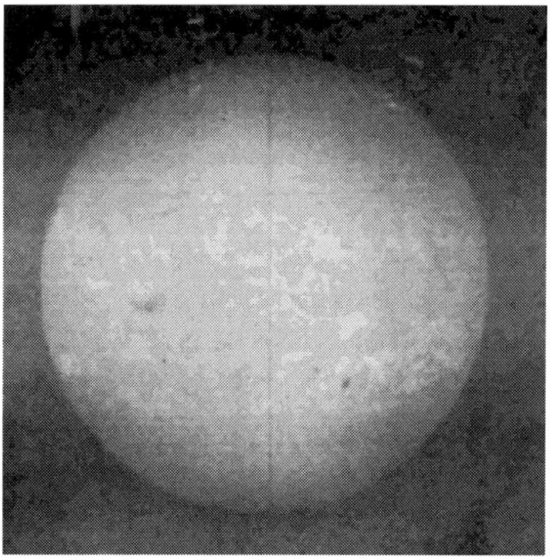

FIG. 73.— ***Spectroheliograph, showing distribution of faculæ upon the sun.—*** *HALE.*

122. ***Faculæ.***— We have seen in Fig. 69 a few of the bright spots called faculæ. At the telescope or in the ordinary photograph these can be seen only at the edge of the sun, because elsewhere the background furnished by the photosphere is so bright that they are lost in it. It is possible, however, by an ingenious application of the spectroscope to break up the sunlight into a spectrum in such a way as to diminish the brightness of this background, much more than the brightness of the faculæ is diminished, and in this way to obtain a photograph of the sun's surface which shall show them wherever they occur, and such a photograph, showing faintly the spectral lines, is reproduced in Fig. 73. The faculæ are the bright patches which stretch inconspicuously across the face of the sun, in two rather irregular belts with a comparatively empty lane between them. This lane lies along the sun's equator, and it is upon either side of it between latitudes 5° and 40° that faculæ seem to be produced. It is significant of their connection with sun

spots that the spots occur in these particular zones and are rarely found outside them.

FIG. 74.— Eclipse of July 20, 1878.— TROUVELOT.

FIG. 75.— Eclipse of April 16, 1893.— SCHAEBERLE.

123. ***Invisible parts of the sun. The Corona.*** — Thus far we have been dealing with parts of the sun that may be seen and photographed under all

ordinary conditions. But outside of and surrounding these parts is an envelope, or rather several envelopes, of much greater extent than the visible sun. These envelopes are for the most part invisible save at those times when the brighter central portions of the sun are hidden in a total eclipse.

FIG. 76.—Eclipse of January 21, 1898.—CAMPBELL.

Fig. 74 is from a drawing, and Figs. 75 and 76 are from eclipse photographs showing this region, in which the most conspicuous object is the halo of soft light called the *corona*, that completely surrounds the sun but is seen to be of differing shapes and differing extent at the several eclipses here shown, although a large part of these apparent differences is due to technical difficulties in photographing, and reproducing an object with outlines so vague as those of the corona. The outline of the corona is so indefinite and its outer portions so faint that it is impossible to assign to it precise dimensions, but at its greatest extent it reaches out for several millions of miles and fills a space more than twenty times as large as the visible part of the sun. Despite its huge bulk, it is of most unsubstantial character, an airy nothing through which comets have been known to force their way around the sun from one side to the other, literally for millions of miles, without having their course influenced or their velocity checked to any appreciable extent. This would hardly be possible if the density even at the bottom of the corona were greater than that of the best vacuum which we are able to produce in laboratory experiments. It seems odd that a vacuum should give off so bright a light as the coronal pictures show, and the exact character of that light and the nature of the corona are still subjects of dispute among astronomers, although it is generally agreed that, in part at least, its light is ordinary sunlight faintly reflected from the widely scattered molecules composing the substance of the corona. It is also probable that in part the light has its origin in the corona itself. A curious and at present unconfirmed result announced by one of the observers of the eclipse of May 28, 1900, is that *the corona is not hot*, its effective temperature being lower than that of the instrument used for the observation.

FIG. 77.— ***Solar prominence of March 25, 1895.*** *— HALE.*

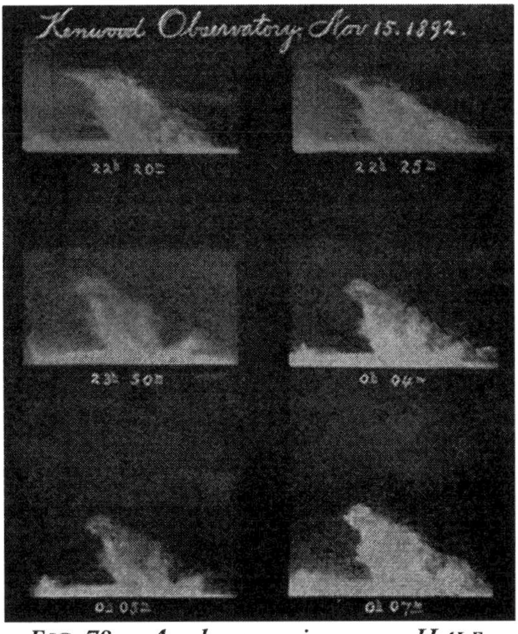

FIG. 78.— ***A solar prominence.*** *— HALE.*

124. ***The chromosphere.***— Between the corona and the photosphere there is a thin separating layer called the *chromosphere* (Greek, color sphere), because when seen at an eclipse it shines with a brilliant red light quite unlike anything else upon the sun save the *prominences* which are themselves only parts of the chromosphere temporarily thrown above its surface, as in a fountain a jet of water is thrown up from the basin and remains for a few moments suspended in mid-air. Not infrequently in such a fountain foreign matter is swept up by the rush of the water— dirt, twigs, small fish, etc.— and in like manner the prominences often carry along with them parts of the underlying layers of the sun, photosphere, faculæ, etc., which reveal their presence in the prominence by adding their characteristic lines to the spectrum, like that of the chromosphere, which the prominence presents when they are absent. None of the eclipse photographs (Figs. 74 to 76) show the chromosphere, because the color effect is lacking in them, but a great curving prominence may be seen near the bottom of Fig. 75, and smaller ones at other parts of the sun's edge.

125. ***Prominences.***— Fig. 77 shows upon a larger scale one of these prominences rising to a height of 160,000 miles above the photosphere; and another photograph, taken 18 minutes later, but not reproduced here, showed the same prominence grown in this brief interval to a stature of 280,000 miles. These pictures were not taken during an eclipse, but in full sunlight, using the same spectroscopic apparatus which was employed in connection with the faculæ to diminish the brightness of the background without much enfeebling the brilliancy of the prominence itself. The dark base from which the prominence seems to spring is not the sun's edge, but a part of the apparatus used to cut off the direct sunlight.

Fig. 78 contains a series of photographs of another prominence taken within an interval of 1 hour 47 minutes and showing changes in size and shape which are much more nearly typical of the ordinary prominence than was the very unusual change in the case of Fig. 77.

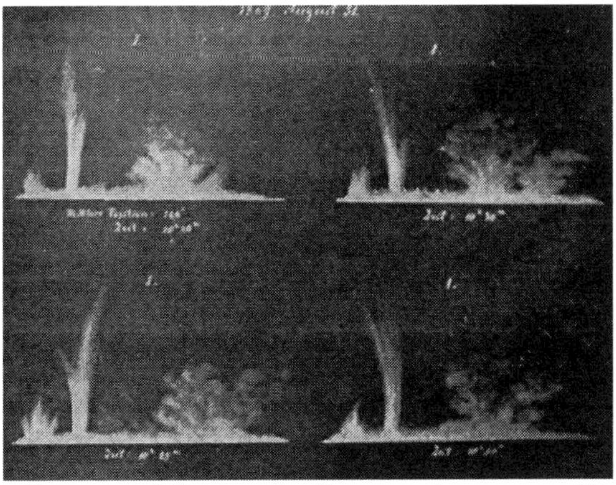

FIG. 79.— Contrasted forms of solar prominences.— ZOELLNER.

The preceding pictures are from photographs, and with them the student may compare Fig. 79, which is constructed from drawings made at the spectroscope by the German astronomer Zoellner. The changes here shown are most marked in the prominence at the left, which is shaped like a broken tree trunk, and which appears to be vibrating from one side to the other like a reed shaken in the wind. Such a prominence is frequently called an *eruptive* one, a name suggested by its appearance of having been blown out from the sun by something like an explosion, while the prominence at the right in this series of drawings, which appears much less agitated, is called by contrast with the other a *quiescent* prominence. These quiescent prominences are, as a rule, much longer-lived than the eruptive ones. One more picture of prominences (Fig. 80) is introduced to show the continuous stretch of chromosphere out of which they spring.

FIG. 80.— Prominences and chromosphere.— HALE.

Prominences are seen only at the edge of the sun, because it is there alone that the necessary background can be obtained, but they must occur at the center of the sun and elsewhere quite as well as at the edge, and it is probable that quiescent prominences are distributed over all parts of the sun's surface, but eruptive prominences show a strong tendency toward the regions of sun spots and faculæ as if all three were intimately related phenomena.

126. ***The sun as a machine.***— Thus far we have considered the anatomy of the sun, dissecting it into its several parts, and our next step should be a consideration of its physiology, the relation of the parts to each other, and their function in carrying on the work of the solar organism, but this step, unfortunately, must be a lame one. The science of astronomy to-day possesses no comprehensive and well-established theory of this kind, but looks to the future for the solution of this the greatest pending problem of solar physics. Progress has been made toward its solution, and among the steps of this progress that we shall have to consider, the first and most important is the conception of the sun as a kind of heat engine.

In a steam engine coal is burned under the boiler, and its chemical energy, transformed into heat, is taken up by the water and delivered, through steam as a medium, to the engine, which again transforms and gives it out as mechanical work in the turning of shafts, the driving of machinery, etc. Now, the function of the sun is exactly opposite to that of the engine and boiler: it gives out, instead of receiving, radiant energy; but, like the engine, it must be fed from some source; it can not be run upon nothing at all any more than the engine can run day after day without fresh supplies of fuel under its boiler. We know that for some thousands of years the sun has been furnishing light and heat to the earth in practically unvarying amount, and not to the earth alone, but it has been pouring forth these forms of energy in every direction, without apparent regard to either use or economy. Of all the radiant energy given off by the sun, only two parts out of every thousand million fall upon any planet of the solar system, and of this small fraction the earth takes about one tenth for the maintenance of its varied forms of life and action. Astronomers and physicists have sought on every hand for an explanation of the means by which this tremendous output of energy is maintained century after century without sensible diminution, and have come with almost one mind to the conclusion that the gravitative forces which reside in the sun's own mass furnish the only adequate explanation for it, although they may be in some small measure re-enforced by minor influences, such as the fall of meteoric dust and stones into the sun.

Every boy who has inflated a bicycle tire with a hand pump knows that the pump grows warm during the operation, on account of the compression of the air within the cylinder. A part of the muscular force (energy) expended in working the pump reappears in the heat which warms both air and pump, and a similar process is forever going on in the sun, only in place of muscular force we must there substitute the tremendous attraction of gravitation, 28 times as great as upon the earth. "The matter in the interior of the sun must be as a shuttlecock between the stupendous pressure and the enormously high temperature," the one tending to compress and the other to expand it, but with this important difference between them: the temperature steadily tends to fall as the heat energy is wasted away, while the gravitative force suffers no corresponding diminution, and in the long run must gain the upper hand, causing the sun to shrink and become more dense. It is this progressive shrinking and compression of its molecules into a smaller space which supplies the energy contained in the sun's output of light and heat. According to Lord Kelvin, each centimeter of shrinkage in the sun's diameter furnishes the energy required to keep up its radiation for something more than an hour, and, on account of the sun's great distance, the shrinkage might go on at this rate for many centuries without producing any measurable effect in the sun's appearance.

127. ***Gaseous constitution of the sun.*** – But Helmholtz's dynamical theory of the maintenance of the sun's heat, which we are here considering, includes one essential feature that is not sufficiently stated above. In order that the explanation may hold true, it is necessary that the sun should be in the main a gaseous body, composed from center to circumference of gases instead of solid or liquid parts. Pumping air warms the bicycle pump in a way that pumping water or oil will not.

The high temperature of the sun itself furnishes sufficient reason for supposing the solar material to be in the gaseous state, but the gas composing those parts of the sun below the photosphere must be very different in some of its characteristics from the air or other gases with which we are familiar at the earth, since its average density is 1,000 times as great as that of air, and its consistence and mechanical behavior must be more like that of honey or tar than that of any gas with which we are familiar. It is worth noting, however, that if a hole were dug into the crust of the earth to a depth of 15 or 20 miles the air at the bottom of the hole would be compressed by that above it to a density comparable with that of the solar gases.

128. ***The sun's circulation.***— It is plain that under the conditions which exist in the sun the outer portions, which can radiate their heat freely into space, must be cooler than the inner central parts, and this difference of temperature must set up currents of hot matter drifting upward and outward from within the sun and counter currents of cooler matter settling down to take its place. So, too, there must be some level at which the free radiation into outer space chills the hot matter sufficiently to condense its less refractory gases into clouds made up of liquid drops, just as on a cloudy day there is a level in our own atmosphere at which the vapor of water condenses into liquid drops which form the thin shell of clouds that hovers above the earth's surface, while above and below is the gaseous atmosphere. In the case of the sun this cloud layer is always present and is that part which we have learned to call the photosphere. Above the photosphere lies the chromosphere, composed of gases less easily liquefied, hydrogen is the chief one, while between photosphere and chromosphere is a thin layer of metallic vapors, perhaps indistinguishable from the top crust of the photosphere itself, which by absorbing the light given off from the liquid photosphere produces the greater part of the Fraunhofer lines in the solar spectrum.

From time to time the hot matter struggling up from below breaks through the photosphere and, carrying with it a certain amount of the metallic vapors, is launched into the upper and cooler regions of the sun, where, parting with its heat, it falls back again upon the photosphere and is absorbed into it. It is altogether probable that the corona is chiefly composed of fine particles ejected from the sun with velocities sufficient to carry them to a height of millions of miles, or even sufficient to carry them off never to return. The matter of the corona must certainly be in a state of the most lively agitation, its particles being alternately hurled up from the photosphere and falling back again like fireworks, the particles which make up the corona of to-day being quite a different set from those of yesterday or last week. It seems beyond question that the prominences and faculæ too are produced in some way by this up-and-down circulation of the sun's matter, and that any mechanical explanation of the sun must be worked out along these lines; but the problem is an exceedingly difficult one, and must include and explain many other features of the sun's activity of which only a few can be considered here.

129. ***The sun-spot period.***– Sun spots come and go, and at best any particular spot is but short-livcd, rarely lasting more than a month or two, and more often its duration is a matter of only a few days. They are not equally numerous at all times, but, like swarms of locusts, they seem to come and abound for a season and then almost to disappear, as if the forces which produced them were of a periodic character alternately active and quiet. The effect of this periodic activity since 1870 is shown in Fig. 81, where the horizontal line is a scale of times, and the distance of the curve above this line for any year shows the relative number of spots which appeared upon the sun in that year. This indicates very plainly that 1870, 1883, and 1893 were years of great sun-spot activity, while 1879 and 1889 were years in which few spots appeared. The older records, covering a period of two centuries, show the same fluctuations in the frequency of sun spots and from these records curves (which may be found in Young's, The Sun) have been plotted, showing a succession of waves extending back for many years.

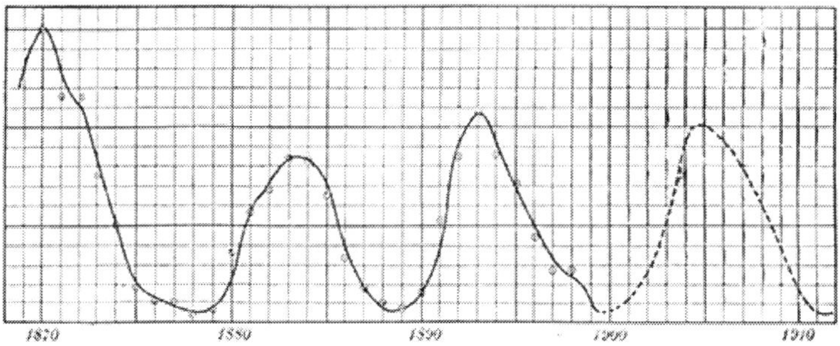

FIG. 81.— The curve of sun-spot frequency.

The sun-spot period is the interval of time from the crest or hollow of one wave to the corresponding part of the next one, and on the average this appears to be a little more than eleven years, but is subject to considerable variation. In accordance with this period there is drawn in broken lines at the right of Fig. 81 a predicted continuation of the sun-spot curve for the first decade of the twentieth century. The irregularity shown by the three preceding waves is such that we must not expect the actual course of future sun spots to correspond very closely to the prediction here made; but in a general way 1901 and 1911 will probably be years of few sun spots, while they will be numerous in 1905, but whether more or less numerous than at preceding epochs of greatest frequency can not be foretold with any approach to certainty so long as we remain in our present ignorance of the causes which make the sun-spot period.

Determine from Fig. 81 as accurately as possible the length of the sun-spot period. It is hard to tell the exact position of a crest or hollow of the curve.

Would it do to draw a horizontal line midway between top and bottom of the curve and determine the length of the period from its intersections with the curve– e. g., in 1874 and 1885?

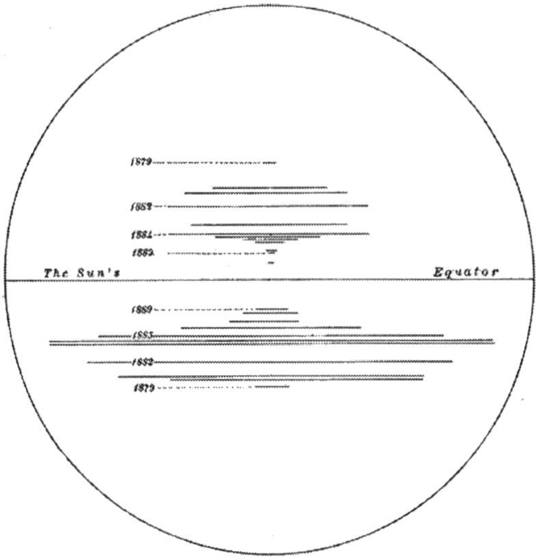

FIG. 82.— Illustrating change of the sun-spot zones.

130. ***The sun-spot zones.***— It has been already noted that sun spots are found only in certain zones of latitude upon the sun, and that faculæ and eruptive prominences abound in these zones more than elsewhere, although not strictly confined to them. We have now to note a peculiarity of these zones which ought to furnish a clew to the sun's mechanism, although up to the present time it has not been successfully traced out. Just before a sun-spot minimum the few spots which appear are for the most part clustered near the sun's equator. As these spots die out two new groups appear, one north the other south of the sun's equator and about 25° or 30° distant from it, and as the period advances toward a maximum these groups shift their positions more and more toward the equator, thus approaching each other but leaving between them a vacant lane, which becomes steadily narrower until at the close of the period, when the next minimum is at hand, it reaches its narrowest dimensions, but does not altogether close up even then. In Fig. 82 these relations are shown for the period falling between 1879 and 1890, by means of the horizontal lines; for each year one line in the northern and one in the southern hemisphere of the sun, their lengths being proportional to the number of spots which appeared in the corresponding hemisphere during the year, and their positions on the sun's disk showing the average latitude of the spots in question. It is very apparent from the figure that during this decade the sun's southern hemisphere was much more active than the northern one in the production of spots, and this appears to be

generally the case, although the difference is not usually as great as in this particular decade.

131. ***Influence of the sun-spot period.***– Sun spots are certainly less hot than the surrounding parts of the sun's surface, and, in view of the intimate dependence of the earth upon the solar radiation, it would be in no way surprising if their presence or absence from the sun's face should make itself felt in some degree upon the earth, raising and lowering its temperature and quite possibly affecting it in other ways. Ingenious men have suggested many such kinds of influence, which, according to their investigations, appear to run in cycles of eleven years. Abundant and scanty harvests, cyclones, tornadoes, epidemics, rainfall, etc., are among these alleged effects, and it is possible that there may be a real connection between any or all of them and the sun-spot period, but for the most part astronomers are inclined to hold that there is only one case in which the evidence is strong enough to really establish a connection of this kind. The magnetic condition of the earth and its disturbances, which are called magnetic storms, do certainly follow in a very marked manner the course of sun-spot activity, and perhaps there should be added to this the statement that auroras (northern lights) stand in close relation to these magnetic disturbances and are most frequent at the times of sun-spot maxima.

Upon the sun, however, the influence of the spot period is not limited to things in and near the photosphere, but extends to the outermost limits of the corona. Determine from Fig. 81 the particular part of the sun-spot period corresponding to the date of each picture of the corona and note how the pictures which were taken near times of sun-spot minima present a general agreement in the shape and extent of the corona, while the pictures taken at a time of maximum activity of the sun spots show a very differently shaped and much smaller corona.

132. ***The law of the sun's rotation.***– We have seen in a previous part of the chapter how the time required by the sun to make a complete rotation upon its axis may be determined from photographs showing the progress of a spot or group of spots across its disk, and we have now to add that when this is done systematically by means of many spots situated in different solar latitudes it leads to a very peculiar and extraordinary result. Each particular parallel of latitude has its own period of rotation different from that of its neighbors on either side, so that there can be no such thing as a fixed geography of the sun's surface. Every part of it is constantly taking up a new position with respect to every other part, much as if the Gulf of Mexico should be south of the United States this year, southeast of it next year, and at the end of a decade should have shifted around to the opposite side of the earth from us. A meridian of longitude drawn down the Mississippi Valley remains always a straight line, or, rather, great circle, upon the surface of the

earth, while Fig. 83 shows what would become of such a meridian drawn through the equatorial parts of the sun's disk. In the first diagram it appears as a straight line running down the middle of the sun's disk. Twenty-five days later, when the same face of the sun comes back into view again, after making a complete revolution about the axis, the equatorial parts will have moved so much faster and farther than those in higher latitudes that the meridian will be warped as in the second diagram, and still more warped after another and another revolution, as shown in the figure.

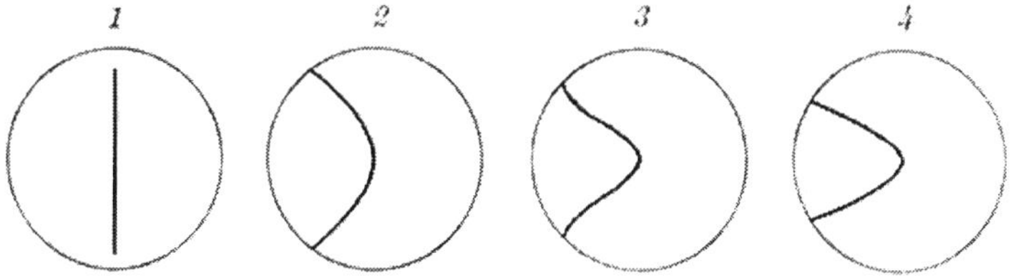

FIG. 83.— Effect of the sun's peculiar rotation in warping a meridian, originally straight.

At least such is the case if the spots truly represent the way in which the sun turns round. There is, however, a possibility that the spots themselves drift with varying speeds across the face of the sun, and that the differences which we find in their rates of motion belong to them rather than to the photosphere. Just what happens in the regions near the poles is hard to say, for the sun spots only extend about halfway from the equator to the poles, and the spectroscope, which may be made to furnish a certain amount of information bearing upon the case, is not as yet altogether conclusive, nor are the faculæ which have also been observed for this purpose.

The simple theory that the solar phenomena are caused by an interchange of hotter and cooler matter between the photosphere and the lower strata of the sun furnishes in its present shape little or no explanation of such features as the sun-spot period, the variations in the corona, the peculiar character of the sun's rotation, etc., and we have still unsolved in the mechanical theory of the sun one of the noblest problems of astronomy, and one upon which both observers and theoretical astronomers are assiduously working at the present time. A close watch is kept upon sun spots and prominences, the corona is observed at every total eclipse, and numerous are the ingenious methods which are being suggested and tried for observing it without an eclipse in ordinary daylight. Attempts, more or less plausible, have been made and are now pending to explain photosphere, spots and the reversing layer by means of the refraction of light within the sun's outer envelope of gases, and it seems altogether probable, in view of these combined activities, that a considerable addition to our store of knowledge concerning the sun may be expected in the not distant future.

CHAPTER X.
THE SUN

CHAPTER XI.
THE PLANETS

133. ***Planets.***— Circling about the sun, under the influence of his attraction, is a family of planets each member of which is, like the moon, a dark body shining by reflected sunlight, and therefore presenting phases; although only two of them, Mercury and Venus, run through the complete series— new, first quarter, full, last quarter— which the moon presents. The way in which their orbits are grouped about the sun has been considered in Chapter III, and Figs. 16 and 17 of that chapter may be completed so as to represent all of the planets by drawing in Fig. 16 two circles with radii of 7.9 and 12.4 centimeters respectively, to represent the orbits of the planets Uranus and Neptune, which are more remote from the sun than Saturn, and by introducing a little inside the orbit of Jupiter about 500 ellipses of different sizes, shapes, and positions to represent a group of minor planets or asteroids as they are often called. It is convenient to regard these asteroids as composing by themselves a class of very small planets, while the remaining 8 larger planets fall naturally into two other classes, a group of medium-sized ones— Mercury, Venus, Earth, and Mars— called inner planets by reason of their nearness to the sun; and the outer planets— Jupiter, Saturn, Uranus, Neptune— each of which is much larger and more massive than any planet of the inner group. Compare in Figs. 84 and 85 their relative sizes. The earth, *E*, is introduced into Fig. 85 as a connecting link between the two figures.

Some of these planets, like the earth, are attended by one or more moons, technically called satellites, which also shine by reflected sunlight and which move about their respective planets in accordance with the law of gravitation, much as the moon moves around the earth.

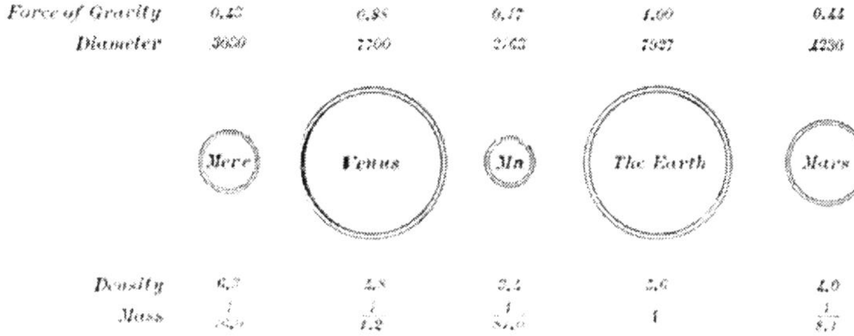

FIG. 84.— The inner planets and the moon.

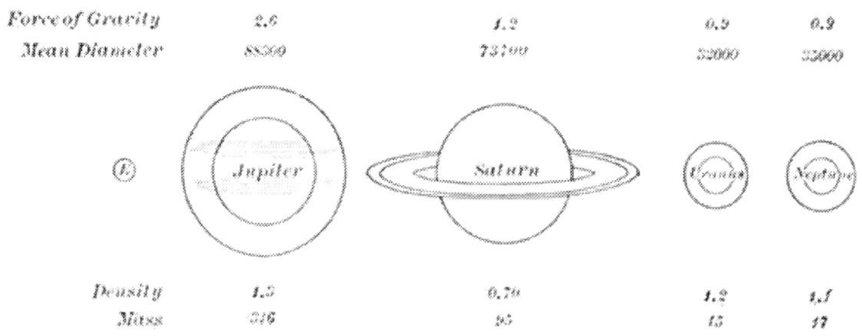

FIG. 85.— The outer planets.

134. ***Distances of the planets from the sun.***— It is a comparatively simple matter to observe these planets year after year as they move among the stars, and to find from these observations how long each one of them requires to make its circuit around the sun— that is, its periodic time, *T*, which figures in Kepler's Third Law, and when these periodic times have been ascertained, to use them in connection with that law to determine the mean distance of each planet from the sun. Thus, Jupiter requires 4,333 days to move completely around its orbit; and comparing this with the periodic time and mean distance of the earth we find—

$a^3 / (4333)^2 = (93{,}000{,}000)^3 / (365.25)^2$,

which when solved gives as the mean distance of Jupiter from the sun, 483,730,000 miles, or 5.20 times as distant as the earth. If we make a similar computation for each planet, we shall find that their distances from the sun show a remarkable agreement with an artificial series of numbers called Bode's law. We write down the numbers contained in the first line of figures below, each of which, after the second, is obtained by doubling the preceding one, add 4 to each number and point off one place of decimals; the resulting number is (approximately) the distance of the corresponding planet from the sun.

Mercury.	Venus .	Earth.	Mars		Jupiter .	Saturn .	Uranus.	Neptune.
0	3	6	12	4	48	96	192	384
4	4	4	4		4	4	4	4
0.4	0.7	1.0	1.6	. 8	5.2	10. 0	19.6	38.8

0.4	0.7	1.0	1.5	. 8	5.2	9.5	19.2	30.1

The last line of figures shows the real distance of the planet as determined from Kepler's law, the earth's mean distance from the sun being taken as the unit for this purpose. With exception of Neptune, the agreement between Bode's law and the true distances is very striking, but most remarkable is the presence in the series of a number, 2.8, with no planet corresponding to it. This led astronomers at the time Bode published the law, something more than a century ago, to give new heed to a suggestion made long before by Kepler, that there might be an unknown planet moving between the orbits of Mars and Jupiter, and a number of them agreed to search for such a planet, each in a part of the sky assigned him for that purpose. But they were anticipated by Piazzi, an Italian, who found the new planet, by accident, on the first day of the nineteenth century, moving at a distance from the sun represented by the number 2.77.

This planet was the first of the asteroids, and in the century that has elapsed hundreds of them have been discovered, while at the present time no year passes by without several more being added to the number. While some of these are nearer to the sun than is the first one discovered, and others are farther from it, their average distance is fairly represented by the number 2.8.

Why Bode's law should hold true, or even so nearly true as it does, is an unexplained riddle, and many astronomers are inclined to call it no law at all, but only a chance coincidence– an illustration of the "inherent capacity of figures to be juggled with"; but if so, it is passing strange that it should represent the distance of the asteroids and of Uranus, which was also an undiscovered planet at the time the law was published.

135. ***The planets compared with each other.*** – When we pass from general considerations to a study of the individual peculiarities of the planets, we find great differences in the extent of knowledge concerning them, and the reason for this is not far to seek. Neptune and Uranus, at the outskirts of the solar system, are so remote from us and so feebly illumined by the sun that any detailed study of them can go but little beyond determining the numbers which represent their size, mass, density, the character of their orbits, etc. The asteroids are so small that in the telescope they look like mere points of light, absolutely indistinguishable in appearance from the fainter stars. Mercury, although closer at hand and presenting a disk of considerable size, always stands so near the sun that its observation is difficult on this account. Something of the same kind is true for Venus, although in much less degree; while Mars, Jupiter, and Saturn are comparatively easy objects for telescopic study, and our knowledge of them, while far from complete, is considerably greater than for the other planets.

Figs. 84 and 85 show the relative sizes of the planets composing the inner and outer groups respectively, and furnish the numerical data concerning their diameters, masses, densities, etc., which are of most importance in judging of their physical condition. Each planet, save Saturn, is represented by two circles, of which the outer is drawn proportional to the size of the planet, and the inner shows the amount of material that must be subtracted from the interior in order that the remaining shell shall just float in water. Note the great difference in thickness of shell between the two groups. Saturn, having a mean density less than that of water, must have something loaded upon it, instead of removed, in order that it should float just submerged.

Jupiter

136. ***Appearance.***— Commencing our consideration of the individual planets with Jupiter, which is by far the largest of them, exceeding both in bulk and mass all the others combined, we have in Fig. 86 four representations of Jupiter and his family of satellites as they may be seen in a very small telescope— e. g., an opera glass— save that the little dots which here represent the satellites are numbered *1*, *2*, *3*, *4*, in order to preserve their identity in the successive pictures.

The chief interest of these pictures lies in the satellites, but, reserving them for future consideration, we note that the planet itself resembles in shape the full moon, although in respect of brightness it sends to us less than 1/6000 part as much light as the moon. From a consideration of the motion of Jupiter and the earth in Fig. 16, show that Jupiter can not present any such phases as does the moon, but that its disk must be at all times nearly full. As seen from Saturn, what kind of phases would Jupiter present?

137. ***The belts.***— Even upon the small scale of Fig. 86 we detect the most characteristic feature of Jupiter's appearance in the telescope, the two bands extending across his face parallel to the line of the satellites, and in Fig. 87 these same dark bands may be recognized amid the abundance of detail which is here brought out by a large telescope. Photography does not succeed as a means of reproducing this detail, and for it we have to rely upon the skill of the artist astronomer. The lettering shows the Pacific Standard time at which the sketches were made, and also the longitude of the meridian of Jupiter passing down the center of the planet's disk.

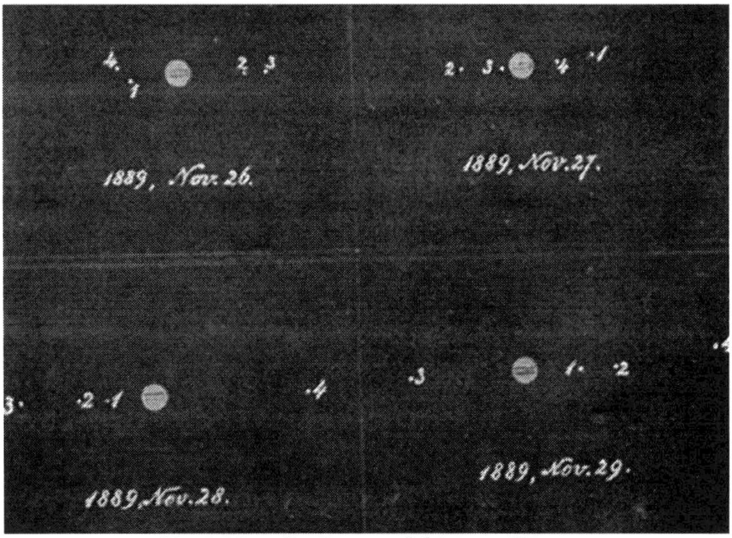

FIG. 86.—Jupiter and his satellites.

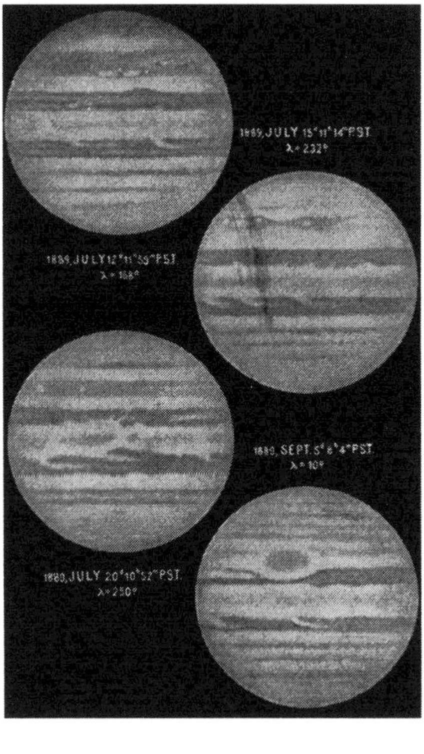

FIG. 87.—Drawings of Jupiter made at the 36-inch telescope of the Lick Observatory.—KEELER.

The dark bands are called technically the belts of Jupiter; and a comparison of these belts in the second and third pictures of the group, in which nearly the same face of the planet is turned toward us, will show that they are subject to considerable changes of form and position even within the space of a few days. So, too, by a comparison of such markings as the round white spots in the upper parts of the disks, and the indentations in the edges of the belts, we may recognize that the planet is in the act of turning round, and must therefore have an axis about which it turns, and poles, an equator, etc. The belts are in fact parallel to the planet's equator; and generalizing from what appears in the pictures, we may say that there is always a strongly marked belt on each side of the equator with a lighter colored streak between them, and that farther from the equator are other belts variable in number, less conspicuous, and less permanent than the two first seen. Compare the position of the principal belts with the position of the zones of sun-spot activity in the sun. A feature of the planet's surface, which can not be here reproduced, is the rich color effect to be found upon it. The principal belts are a brick-red or salmon color, the intervening spaces in general white but richly mottled, and streaked with purples, browns, and greens.

The drawings show the planet as it appeared in the telescope, inverted, and they must be turned upside down if we wish the points of the compass to appear as upon a terrestrial map. Bearing this in mind, note in the last picture the great oval spot in the southern hemisphere of Jupiter. This is a famous marking, known from its color as the *great red spot*, which appeared first in 1878 and has persisted to the present day (1900), sometimes the most conspicuous marking on the planet, at others reduced to a mere ghost of itself, almost invisible save for the indentation which it makes in the southern edge of the belt near it.

138. ***Rotation and flattening at the poles.*** – One further significant fact with respect to Jupiter may be obtained from a careful measurement of the drawings; the planet is flattened at the poles, so that its polar diameter is about one sixteenth part shorter than the equatorial diameter. The flattening of the earth amounts to only one three-hundredth part, and the marked difference between these two numbers finds its explanation in the greater swiftness of Jupiter's rotation about its axis, since in both cases it is this rotation which makes the flattening.

It is not easy to determine the precise dimensions of the planet, since this involves a knowledge both of its distance from us and of the angle subtended by its diameter, but the most recent determinations of this kind assign as the equatorial diameter 90,200 miles, and for the polar diameter 84,400 miles. Determine from either of these numbers the size of the great red spot.

The earth turns on its axis once in 24 hours but no such definite time can be assigned to Jupiter, which, like the sun, seems to have different rotation periods in different latitudes— 9h. 50m. in the equatorial belt and 9h. 56m. in the dark belts and higher latitudes. There is some indication that the larger part of the visible surface rotates in 9h. 55.6m., while a broad stream along the equator flows eastward some 270 miles per hour, and thus comes back to the center of the planet, as seen from the earth, five or six minutes earlier than the parts which do not share in this motion. Judged by terrestrial standards, 270 miles per hour is a great velocity, but Jupiter is constructed on a colossal scale, and, too, we have to compare this movement, not to a current flowing in the ocean, but to a wind blowing in the upper regions of the earth's atmosphere. The visible surface of Jupiter is only the top of a cloud formation, and contains nothing solid or permanent, if indeed there is anything solid even at the core of the planet. The great red spot during the first dozen years of its existence, instead of remaining fixed relative to the surrounding formations, drifted two thirds of the way around the planet, and having come to a standstill about 1891, it is now slowly retracing its path.

139. ***Physical condition.***— For a better understanding of the physical condition of Jupiter, we have now to consider some independent lines of evidence which agree in pointing to the conclusion that Jupiter, although classed with the earth as a planet, is in its essential character much more like the sun.

Appearance.— The formations which we see in Fig. 87 look like clouds. They gather and disappear, and the only element of permanence about them is their tendency to group themselves along zones of latitude. If we measure the light reflected from the planet we find that its albedo is very high, like that of snow or our own cumulus clouds, and it is of course greater from the light parts of the disk than from the darker bands. The spectroscope shows that the sunlight reflected from these darker belts is like that reflected from the lighter parts, save that a larger portion of the blue and violet rays has been absorbed out of it, thus producing the ruddy tint of the belts, as sunset colors are produced on the earth, and showing that here the light has penetrated farther into the planet's atmosphere before being thrown back by reflection from lower-lying cloud surfaces. The dark bands are therefore to be regarded as rifts in the clouds, reaching down to some considerable distance and indicating an atmosphere of great depth. The great red spot, 28,000 miles long, and obviously thrusting back the white clouds on every side of it, year after year, can hardly be a mere patch on the face of the planet, but indicates some considerable depth of atmosphere.

Density.— So, too, the small mean density of the planet, only 1.3 times that of water and actually less than the density of the sun, suggests that the larger part of the planet's bulk may be made of gases and clouds, with very little solid matter even at the center; but here we get into a difficulty from which there seems but one escape. The force of gravity at the visible surface of Jupiter may be found from its mass and dimensions to be 2.6 times as great as at the surface of the earth, and the pressure exerted upon its atmosphere by this force ought to compress the lower strata into something more dense than we find in the planet. Some idea of this compression may be obtained from Fig. 88, where the line marked *E* shows approximately how the density of the air increases as we move from its upper strata down toward the surface of the earth through a distance of 16 miles, the density at any level being proportional to the distance of the curved line from the straight one near it. The line marked *J* in the same figure shows how the density would increase if the force of gravity were as great here as it is in Jupiter, and indicates a much greater rate of increase. Starting from the upper surface of the cloud in Jupiter's atmosphere, if we descend, not 16 miles, but 1,600 or 16,000, what must the density of the atmosphere become and how is this to be reconciled with what we know to be the very small mean density of the planet?

We are here in a dilemma between density on the one hand and the effects of gravity on the other, and the only escape from it lies in the assumption that the interior of Jupiter is tremendously hot, and that this heat expands the substance of the planet in spite of the pressure to which it is subject, making a large planet with a low density, possibly gaseous at the very center, but in its outer part surrounded by a shell of clouds condensed from the gases by radiating their heat into the cold of outer space.

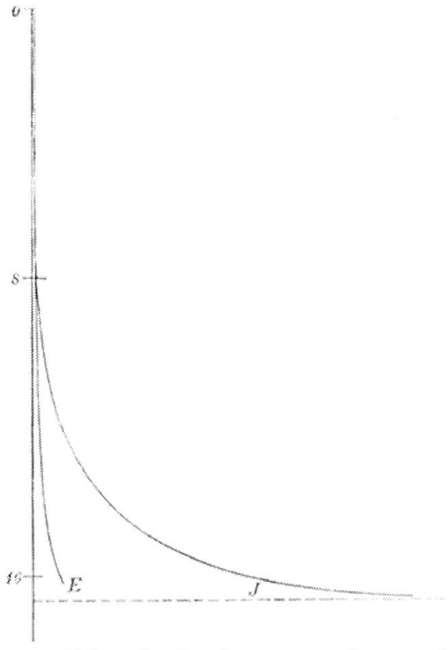

FIG. 88.— Increase of density in the atmospheres of Jupiter and the earth.

This is essentially the same physical condition that we found for the sun, and we may add, as further points of resemblance between it and Jupiter, that there seems to be a circulation of matter from the hot interior of the planet to its cooler surface that is more pronounced in the southern hemisphere than in the northern, and that has its periods of maximum and minimum activity, which, curiously enough, seem to coincide with periods of maximum and minimum sun-spot development. Of this, however, we can not be entirely sure, since it is only in recent years that it has been studied with sufficient care, and further observations are required to show whether the agreement is something more than an accidental and short-lived coincidence.

Temperature.– The temperature of Jupiter must, of course, be much lower than that of the sun, since the surface which we see is not luminous like the sun's; but below the clouds it is not improbable that Jupiter may be incandescent, white hot, and it is surmised with some show of probability that a little of its light escapes through the clouds from time to time, and helps to produce the striking brilliancy with which this planet shines.

140. ***The satellites of Jupiter.***– The satellites bear much the same relation to Jupiter that the moon bears to the earth, revolving about the planet in accordance with the law of gravitation, and conforming to Kepler's three laws, as do the planets in their courses about the sun. Observe in Fig. 86 the position of satellite No. *1* on the four dates, and note how it oscillates back and forth from left to right of Jupiter, apparently making a complete revolution in about two days, while No. *4* moves steadily from left to right during the entire period, and has evidently made only a fraction of a revolution in the time covered by the pictures. This quicker motion, of course, means that No. *1* is nearer to Jupiter than No. *4*, and the numbers given to the satellites show the order of their distances from the planet. The peculiar way in which the satellites are grouped, always standing nearly in a straight line, shows that their orbits must lie nearly in the same plane, and that this plane, which is also the plane of the planets' equator, is turned edgewise toward the earth.

These satellites enjoy the distinction of being the first objects ever discovered with the telescope, having been found by Galileo almost immediately after its invention, A. D. 1610. It is quite possible that before this time they may have been seen with the naked eye, for in more recent years reports are current that they have been seen under favorable circumstances by sharp-eyed persons, and very little telescopic aid is required to show them. Look for them with an opera or field glass. They bear the names Io, Europa, Ganymede, Callisto, which, however, are rarely used, and, following the custom of astronomers, we shall designate them by the Roman numerals I, II, III, IV.

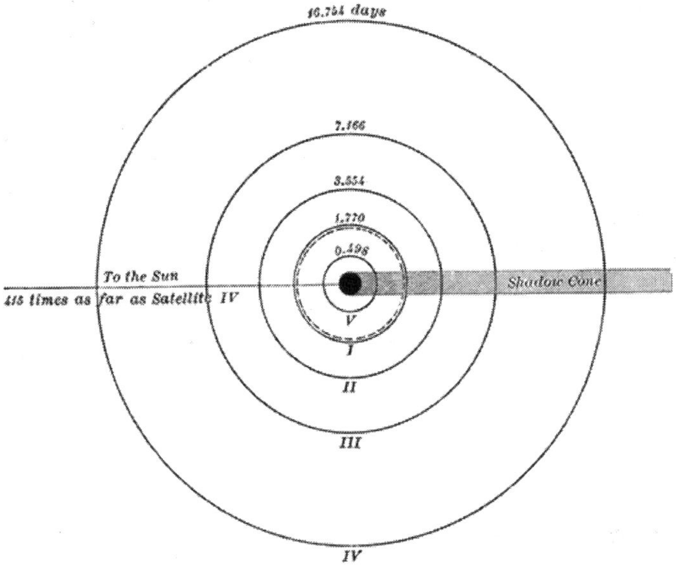

FIG. 89.—Orbits of Jupiter's satellites.

For nearly three centuries (1610 to 1892) astronomers spoke of the four satellites of Jupiter; but in September, 1892, a fifth one was added to the number by Professor Barnard, who, observing with the largest telescope then extant, found very close to Jupiter a tiny object only 1/600 part as bright as the other satellites, but, like them, revolving around Jupiter, a permanent member of his system. This is called the fifth satellite, and Fig. 89 shows the orbits of these satellites around Jupiter, which is here represented on the same scale as the orbits themselves. The broken line just inside the orbit of I represents the size of the moon's orbit. The cut shows also the periodic times of the satellites expressed in days, and furnishes in this respect a striking illustration of the great mass of Jupiter. Satellite I is a little farther from Jupiter than is the moon from the earth, but under the influence of a greater attraction it makes the circuit of its orbit in 1.77 days, instead of taking 29.53 days, as does the moon. Determine from the figure by the method employed in § 111 how much more massive is Jupiter than the earth.

Small as these satellites seem in Fig. 86, they are really bodies of considerable size, as appears from Fig. 90, where their dimensions are compared with those of the earth and moon, save that the fifth satellite is not included. This one is so small as to escape all attempts at measuring its diameter, but, judging from the amount of light it reflects, the period printed with the legend of the figure represents a gross exaggeration of this satellite's size.

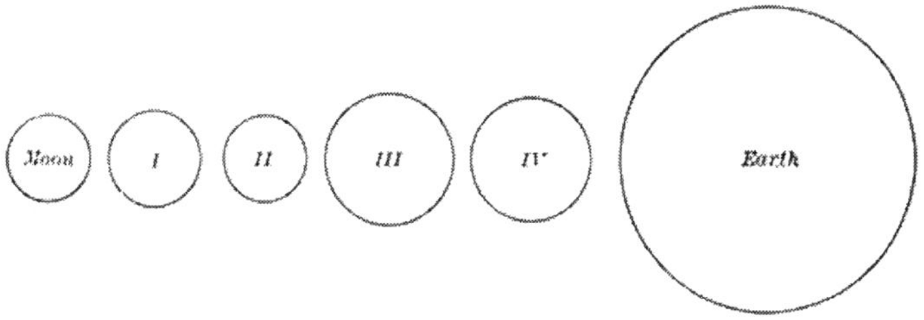

FIG. 90.—Jupiter's satellites compared with the earth and moon.

Like the moon, each of these satellites may fairly be considered a world in itself, and as such a fitting object of detailed study, but, unfortunately, their great distance from us makes it impossible, even with the most powerful telescope, to see more upon their surfaces than occasional vague markings, which hardly suffice to show the rotations of the satellites upon their axes.

One striking feature, however, comes out from a study of their influence in disturbing each other's motion about Jupiter. Their masses and the resulting densities of the satellites are smaller than we should have expected to find, the density being less than that of the moon, and averaging only a little greater than the density of Jupiter itself. At the surface of the third satellite the force of gravity is but little less than on the moon, although the moon's density is nearly twice as great as that of III, and there can be no question here of accounting for the low density through expansion by great heat, as in the case of the sun and Jupiter. It has been surmised that these satellites are not solid bodies, like the earth and moon, but only shoals of rock and stone, loosely piled together and kept from packing into a solid mass by the action of Jupiter in raising tides within them. But the explanation can hardly be regarded as an accepted article of astronomical belief, although it is supported by some observations which tend to show that the apparent shapes of the satellites change under the influence of the tidal forces impressed upon them.

141. ***Eclipses of the satellites.***— It may be seen from Fig. 89 that in their motion around the planet Jupiter's satellites must from time to time pass through his shadow and be eclipsed, and that the shadows of the satellites will occasionally fall upon the planet, producing to an observer upon Jupiter an eclipse of the sun, but to an observer on the earth presenting only the appearance of a round black spot moving slowly across the face of the planet. Occasionally also a satellite will pass exactly between the earth and Jupiter, and may be seen projected against the planet as a background. All of these phenomena are duly predicted and observed by astronomers, but the eclipses are the only ones we need consider here. The importance of these eclipses was early recognized, and astronomers endeavored to construct a theory of their recurrence which would permit accurate predictions of them to be made. But in this they met with no great success, for while it was easy enough to

foretell on what night an eclipse of a given satellite would occur, and even to assign the hour of the night, it was not possible to make the predicted minute agree with the actual time of eclipse until after Roemer, a Danish astronomer of the seventeenth century, found where lay the trouble. His discovery was, that whenever the earth was on the side of its orbit toward Jupiter the eclipses really occurred before the predicted time, and when the earth was on the far side of its orbit they came a few minutes later than the predicted time. He correctly inferred that this was to be explained, not by any influence which the earth exerted upon Jupiter and his satellites, but through the fact that the light by which we see the satellite and its eclipse requires an appreciable time to cross the intervening space, and a longer time when the earth is far from Jupiter than when it is near.

For half a century Roemer's views found little credence, but we know now that he was right, and that on the average the eclipses come 8m. 18s. early when the earth is nearest to Jupiter, and 8m. 18s. late when it is on the opposite side of its orbit. This is equivalent to saying that light takes 8m. 18s. to cover the distance from the sun to the earth, so that at any moment we see the sun not as it then is, but as it was 8 minutes earlier. It has been found possible in recent years to measure by direct experiment the velocity with which light travels— 186,337 miles per second— and multiplying this number by the 498s. (= 8m. 18s.) we obtain a new determination of the sun's distance from the earth. The product of the two numbers is 92,795,826, in very fair agreement with the 93,000,000 miles found in Chapter X; but, as noted there, this method, like every other, has its weak side, and the result may be a good many thousands of miles in error.

It is worthy of note in this connection that both methods of obtaining the sun's distance which were given in Chapter X involve Kepler's Third Law, while the result obtained from Jupiter's satellites is entirely independent of this law, and the agreement of the several results is therefore good evidence both for the truth of Kepler's laws and for the soundness of Roemer's explanation of the eclipses. This mode of proof, by comparing the numerical results furnished by two or more different principles, and showing that they agree or disagree, is of wide application and great importance in physical science.

Saturn

FIG. 91.—Aspects of Saturn's rings.

142. ***The ring of Saturn.***— In respect of size and mass Saturn stands next to Jupiter, and although far inferior to him in these respects, it contains more material than all the remaining planets combined. But the unique feature of Saturn which distinguishes it from every other known body in the heavens is its ring, which was long a puzzle to the astronomers who first studied the planet with a telescope (one of them called Saturn a planet with ears), but, was after nearly half a century correctly understood and described by Huyghens, whose Latin text we translate into— "It is surrounded by a ring, thin, flat, nowhere touching it, and making quite an angle with the ecliptic."

Compare with this description Fig. 91, which shows some of the appearances presented by the ring at different positions of Saturn in its orbit. It was their varying aspects that led Huyghens to insert the last words of his description, for, if the plane of the ring coincided with the plane of the earth's orbit, then at all times the ring must be turned edgewise toward the earth, as shown in the middle picture of the group. Fig. 92 shows the sun and the orbit of the earth placed near the center of Saturn's orbit, across whose circumference are ruled some oblique lines representing the plane of the ring, the right end always tilted up, no matter where the planet is in its orbit. It is evident that an observer upon the earth will see the *N* side of the ring when the planet is at *N* and the *S* side when it is at *S*, as is shown in the first and third pictures of Fig. 91, while midway between these positions the edge of the ring will be presented to the earth.

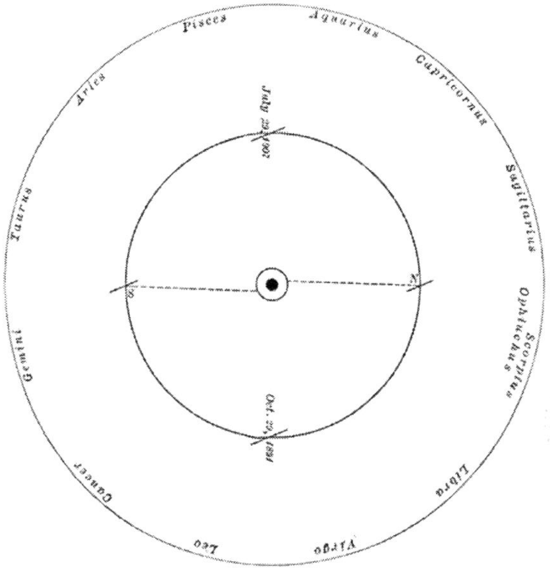

FIG. 92.—Aspects of the ring in their relation to Saturn's orbital motion.

The last occasion of this kind was in October, 1891, and with the large telescope of the Washburn Observatory the writer at that time saw Saturn without a trace of a ring surrounding it. The ring is so thin that it disappears altogether when turned edgewise. The names of the zodiacal constellations are inserted in Fig. 92 in their proper direction from the sun, and from these we learn that the ring will disappear, or be exceedingly narrow, whenever Saturn is in the constellation Pisces or near the boundary line between Leo and Virgo. It will be broad and show its northern side when Saturn is in Scorpius or Sagittarius, and its southern face when the planet is in Gemini. What will be its appearance in 1907 at the date marked in the figure?

143. ***Nature of the ring.***— It is apparent from Figs. 91 and 93 that Saturn's ring is really made up of two or more rings lying one inside of the other and completely separated by a dark space which, though narrow, is as clean and sharp as if cut with a knife. Also, the inner edge of the ring fades off into an obscure border called the *dusky ring* or *crape ring*. This requires a pretty good telescope to show it, as may be inferred from the fact that it escaped notice for more than two centuries during which the planet was assiduously studied with telescopes, and was discovered at the Harvard College Observatory as recently as 1850.

Although the rings appear oval in all of the pictures, this is mainly an effect of perspective, and they are in fact nearly circular with the planet at their center. The extreme diameter of the ring is 172,000 miles, and from this number, by methods already explained (Chapter IX), the student should obtain the width of the rings, their distance from the ball of the planet, and the diameter of the ball. As to thickness, it is evident, from the disappearance of the ring when its edge is turned toward the earth, that it is very thin in comparison with its diameter, probably not more than 100 miles thick, although no exact measurement of this can be made.

FIG. 93.— Saturn.

From theoretical reasons based upon the law of gravitation astronomers have held that the rings of Saturn could not possibly be solid or liquid bodies. The strains impressed upon them by the planet's attraction would tear into fragments steel rings made after their size and shape. Quite recently Professor Keeler has shown, by applying the spectroscope (Doppler's principle) to determine the velocity of the ring's rotation about Saturn, that the inner parts of the ring move, as Kepler's Third Law requires, more rapidly than do the outer parts, thus furnishing a direct proof that they are not solid, and leaving no doubt that they are made up of separate fragments, each moving about the planet in its own orbit, like an independent satellite, but standing so close to its neighbors that the whole space reflects the sunlight as completely as if it were solid. With this understanding of the rings it is easy to see why they are so thin. Like Jupiter, Saturn is greatly flattened at the poles, and this flattening, or rather the protuberant mass about the equator, lays hold of every satellite near the planet and exerts upon it a direct force tending to thrust it down into the plane of the planet's equator and hold it there. The ring lies in the plane of Saturn's equator because each particle is constrained to move there.

The division of the ring into two parts, an outer and an inner ring, is usually explained as follows: Saturn is surrounded by a numerous brood of satellites, which by their attractions produce perturbations in the material composing the rings, and the dividing line between the outer and inner rings falls at the place where by the law of gravitation the perturbations would have their greatest effect. The dividing line between the rings is therefore a narrow lane, 2,400 miles wide, from which the fragments have been swept clean away by the perturbing action of the satellites. Less conspicuous divisions are seen from time to time in other parts of the ring, where the perturbations, though less, are still appreciable. But it is open to some question whether this explanation is sufficient.

The curious darkness of the inner or crape ring is easily explained. The particles composing it are not packed together so closely as in the outer ring, and therefore reflect less sunlight. Indeed, so sparsely strewn are the particles in this ring that it is in great measure transparent to the sunlight, as is shown by a recorded observation of one of the satellites which was distinctly although faintly seen while moving through the shadow of the dark ring, but disappeared in total eclipse when it entered the shadow cast by the bright ring.

144. ***The ball of Saturn.*** – The ball of the planet is in most respects a smaller copy of Jupiter. With an equatorial diameter of 76,000 miles, a polar diameter of 69,000 miles, and a mass 95 times that of the earth, its density is found to be the least of any planet in the solar system, only 0.70 of the density of water, and about one half as great as is the density of Jupiter. The force of gravity at its surface is only a little greater (1.18) than on the earth; and this, in connection with the low density, leads, as in the case of Jupiter, to the conclusion that the planet must be mainly composed of gases and vapors, very hot within, but inclosed by a shell of clouds which cuts off their glow from our eyes.

Like Jupiter in another respect, the planet turns very swiftly upon its axis, making a revolution in 10 hours 14 minutes, but up to the present it remains unknown whether different parts of the surface have different rotation times.

145. ***The satellites.*** – Saturn is attended by a family of nine satellites, a larger number than belongs to any other planet, but with one exception they are exceedingly small and difficult to observe save with a very large telescope. Indeed, the latest one is said to have been discovered in 1898 by means of the

image which it impressed upon a photographic plate, and it has never been *seen*.

Titan, the largest of them, is distant 771,000 miles from the planet and bears much the same relation to Saturn that Satellite III bears to Jupiter, the similarity in distance, size and mass being rather striking, although, of course, the smaller mass of Saturn as compared with Jupiter makes the periodic time of Titan— 15 days 23 hours— much greater than that of III. Can you apply Kepler's Third Law to the motion of Titan so as to determine from the data given above, the time required for a particle at the outer or inner edge of the ring to revolve once around Saturn?

Japetus, the second satellite in point of size, whose distance from Saturn is about ten times as great as the moon's distance from the earth, presents the remarkable peculiarity of being always brighter in one part of its orbit than in another, three or four times as bright when west of Saturn as when east of it. This probably indicates that, like our own moon, the satellite turns always the same face toward its planet, and further, that one side of the satellite reflects the sunlight much better than the other side— i. e., has a higher albedo. With these two assumptions it is easily seen that the satellite will always turn toward the earth one face when west, and the other face when east of Saturn, and thus give the observed difference of brightness.

Uranus and Neptune

146. ***Chief characteristics.***— The two remaining large planets are interesting chiefly as modern additions to the known members of the sun's family. The circumstances leading to the discovery of Neptune have been touched upon in Chapter IV, and for Uranus we need only note that it was found by accident in the year 1781 by William Herschel, who for some time after the discovery considered it to be only a comet. It was the first planet ever discovered, all of its predecessors having been known from prehistoric times.

William Herschel (1738-1822).

Uranus has four satellites, all of them very faint, which present only one feature of special importance. Instead of moving in orbits which are approximately parallel to the plane of the ecliptic, as do the satellites of the inner planets, their orbit planes are tipped up nearly perpendicular to the planes of the orbits of both Uranus and the earth. The one satellite which Neptune possesses has the same peculiarity in even greater degree, for its motion around the planet takes place in the direction opposite to that in which all the planets move around the sun, much as if the orbit of the satellite had been tipped over through an angle of 150°. Turn a watch face down and note how the hands go round in the direction opposite to that in which they moved before the face was turned through 180°.

Both Uranus and Neptune are too distant to allow much detail to be seen upon their surfaces, but the presence of broad absorption bands in their spectra shows that they must possess dense atmospheres quite different in constitution from the atmosphere of the earth. In respect of density and the force of gravity at their surfaces, they are not very unlike Saturn, although their density is greater and gravity less than his, leading to the supposition that they are for the most part gaseous bodies, but cooler and probably more nearly solid than either Jupiter or Saturn.

Under favorable circumstances Uranus may be seen with the naked eye by one who knows just where to look for it. Neptune is never visible save in a telescope.

147. ***The inner planets.***— In sharp contrast with the giant planets which we have been considering stands the group of four inner planets, or five if we count the moon as an independent body, which resemble each other in being all small, dense, and solid bodies, which by comparison with the great distances separating the outer planets may fairly be described as huddled together close to the sun. Their relative sizes are shown in Fig. 84, together with the numerical data concerning size, mass, density, etc., which we have already found important for the understanding of a planet's physical condition.

Venus

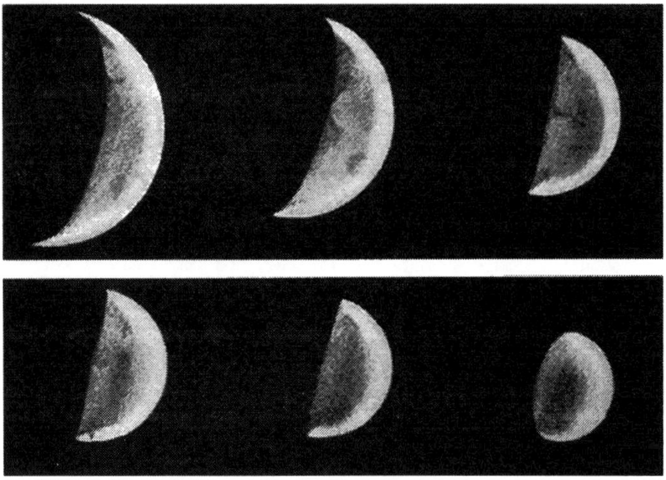

FIG. 94.— The phases of Venus.— ANTONIADI.

148. ***Appearance.***— Omitting the earth, Venus is by far the most conspicuous member of this group, and when at its brightest is, with exception of the sun and moon, the most brilliant object in the sky, and may be seen with the naked eye in broad daylight if the observer knows just where to look for it. But its brilliancy is subject to considerable variations on account of its changing distance from the earth, and the apparent size of its disk varies for the same reason, as may be seen from Fig. 94. These drawings bring out well the phases of the planet, and the student should determine from Fig. 17 what are the relative positions in their orbits of the earth and Venus at which the planet would present each of these phases. As a guide to this, observe that the dark part of Venus's earthward side is always proportional in area to the angle at Venus between the earth and sun. In the first picture of Fig. 94 about two thirds of the surface corresponding to the full hemisphere of the planet is dark, and the angle at Venus between earth and sun is therefore two thirds of 180°— i. e., 120°. In Fig. 17 find a place on the orbit of Venus from which if lines be drawn to the sun and earth, as there shown, the angle between them will be 120°. Make a similar construction for the fourth picture in Fig. 94. Which of these two positions is farther from the earth? How do the distances compare with the apparent size of Venus in the two pictures? What is the phase of Venus to-day?

The irregularities in the shading of the illuminated parts of the disk are too conspicuous in Fig. 94, on account of difficulties of reproduction; these shadings are at the best hard to see in the telescope, and distinct permanent markings upon the planet are wholly lacking. This absence of markings makes almost impossible a determination of the planet's time of rotation about its axis, and astronomers are divided in this respect into two parties, one of which maintains that Venus, like the earth, turns upon its axis in some period not very different from 24 hours, while the other contends that, like the moon, it turns always the same face toward the center of its orbit, making a rotation upon its axis in the same period in which it makes a revolution about the sun. The reason why no permanent markings are to be seen on this planet is easily found. Like Jupiter and Saturn, its atmosphere is at all times heavily cloud-laden, so that we seldom, if ever, see down to the level of its solid parts. There is, however, no reason here to suppose the interior parts hot and gaseous. It is much more probable that Venus, like the earth, possesses a solid crust whose temperature we should expect to be considerably higher than that of the earth, because Venus is nearer the sun. But the cloud layer in its atmosphere must modify the temperature in some degree, and we have practically no knowledge of the real temperature conditions at the surface of the planet.

It is the clouds of Venus which in great measure are responsible for its marked brilliancy, since they are an excellent medium for reflecting the sunlight, and give to its surface an albedo greater than that of any other planet, although Saturn is nearly equal to it.

Of course, the presence of such cloud formations indicates that Venus is surrounded by a dense atmosphere, and we have independent evidence of this in the shape of its disk when the planet is very nearly between the earth and sun. The illuminated part, from tip to tip of the horns, then stretches more than halfway around the planet's circumference, and shows that a certain amount of light must have been refracted through its atmosphere, thus making the horns of the crescent appear unduly prolonged. This atmosphere is shown by the spectroscope to be not unlike that of the earth, although, possibly, more dense.

Mercury

149. ***Chief characteristics.***— Mercury, on account of its nearness to the sun, is at all times a difficult object to observe, and Copernicus, who spent most of his life in Poland, is said, despite all his efforts, to have gone to his grave without ever seeing it. In our more southern latitude it can usually be seen for about a fortnight at the time of each elongation— i. e., when at its greatest angular distance from the sun— and the student should find from Fig. 16 the time at which the next elongation occurs and look for the planet, shining like a star of the first magnitude, low down in the sky just after sunset or before sunrise, according as the elongation is to the east or west of the sun. When seen in the morning sky the planet grows brighter day after day until it disappears in the sun's rays, while in the evening sky its brilliancy as steadily diminishes until the planet is lost. It should therefore be looked for in the evening as soon as possible after it emerges from the sun's rays.

Mercury, as the smallest of the planets, is best compared with the moon, which it does not greatly surpass in size and which it strongly resembles in other respects. Careful comparisons of the amount of light reflected by the planet in different parts of its orbit show not only that its albedo agrees very closely with that of the moon, but also that its light changes with the varying phase of the planet in almost exactly the same way as the amount of moonlight changes. We may therefore infer that its surface is like that of the moon, a rough and solid one, with few or no clouds hanging over it, and most probably covered with very little or no atmosphere. Like Venus, its rotation period is uncertain, with the balance of probability favoring the view that it rotates upon its axis once in 88 days, and therefore always turns the same face toward the sun.

If such is the case, its climate must be very peculiar: one side roasted in a perpetual day, where the direct heating power of the sun's rays, when the planet is at perihelion, is ten times as great as on the moon, and which six weeks later, when the planet is at its farthest from the sun, has fallen off to less than half of this. On the opposite side of the planet there must reign perpetual night and perpetual cold, mitigated by some slight access of warmth from the day side, and perhaps feebly imitating the rapid change of season which takes place on the day side of the planet. This view, however, takes no account of a possible deviation of the planet's axis from being perpendicular to the plane of its orbit, or of the librations which must be produced by the great eccentricity of the orbit, either of which would complicate without entirely destroying the ideal conditions outlined above.

Mars

150. ***Appearance.***— The one remaining member of the inner group, Mars, has in recent years received more attention than any other planet, and the newspapers and magazines have announced marvelous things concerning it: that it is inhabited by a race of beings superior in intelligence to men; that the work of their hands may be seen upon the face of the planet; that we should endeavor to communicate with them, if indeed they are not already sending messages to us, etc.— all of which is certainly important, if true, but it rests upon a very slender foundation of evidence, a part of which we shall have to consider.

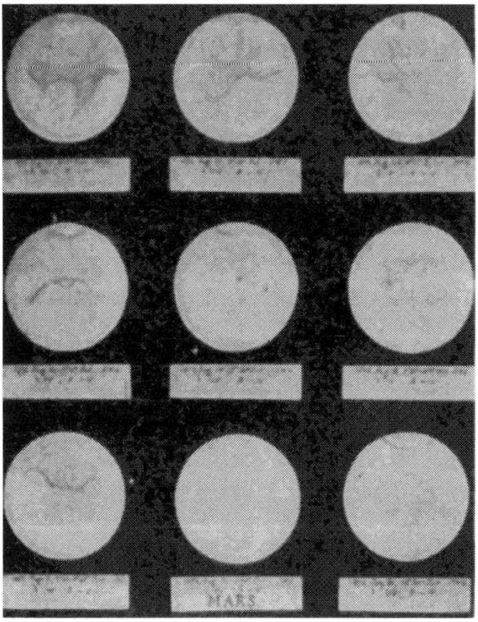

FIG. 95.— Mars.— SCHAEBERLE.

Beginning with facts of which there is no doubt, this ruddy-colored planet, which usually shines about as brightly as a star of the first magnitude, sometimes displays more than tenfold this brilliancy, surpassing every other planet save Venus and presenting at these times especially favorable opportunities for the study of its surface. The explanation of this increase of brilliancy is, of course, that the planet approaches unusually near to the earth, and we have already seen from a consideration of Fig. 17 that this can only happen in the months of August and September. The last favorable epoch of this kind was in 1894. From Fig. 17 the student should determine when the next one will come.

Fig. 95 presents nine drawings of the planet made at one of the epochs of close approach to the earth, and shows that its face bears certain faint markings which, though inconspicuous, are fixed and permanent features of the planet. The dark triangular projection in the lower half of the second drawing was seen and sketched by Huyghens, 1659 A. D. In Fig. 96 some of these markings are shown much more plainly, but Fig. 95 gives a better idea of their usual appearance in the telescope.

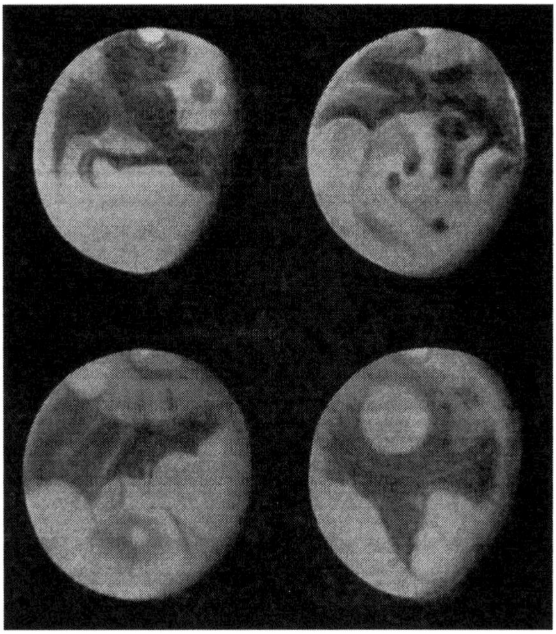

FIG. 96.— Four views of Mars differing 90° in longitude.— BARNARD.

151. ***Rotation.***— It may be seen readily enough, from a comparison of the first two sketches of Fig. 95, that the planet rotates about an axis, and from a more extensive study it is found to be very like the earth in this respect, turning once in 24h. 37m. around an axis tipped from being perpendicular to the plane of its orbit about a degree and a half more than is the earth's axis. Since it is this inclination of the axis which is the cause of changing seasons upon the earth, there must be similar changes, winter and summer, as well as day and night, upon Mars, only each season is longer there than here in the same proportion that its year is longer than ours— i e., nearly two to one. It is summer in the northern hemisphere of Mars whenever the sun, as seen from Mars, stands in that constellation which is nearest the point of the sky toward which the planet's axis points. But this axis points toward the constellation Cygnus, and Alpha Cygni is the bright star nearest the north pole of Mars. As Pisces is the zodiacal constellation nearest to Cygnus, it must be summer in the northern hemisphere of Mars when the sun is in Pisces, or, turning the proposition about, it must be summer in the *southern* hemisphere of Mars when the planet, as seen from the sun, lies in the direction of Pisces.

152. ***The polar caps.***— One effect of the changing seasons upon Mars is shown in Fig. 97, where we have a series of drawings of the region about its south pole made in 1894, on dates between May 21st and December 10th. Show from Fig. 17 that during this time it was summer in the region here

shown. Mars crossed the prime radius in 1894 on September 5th. The striking thing in these pictures is the white spot surrounding the pole, which shrinks in size from the beginning to near the end of the series, and then disappears altogether. The spot came back again a year later, and like a similar spot at the north pole of the planet it waxes in the winter and wanes during the summer of Mars in endless succession.

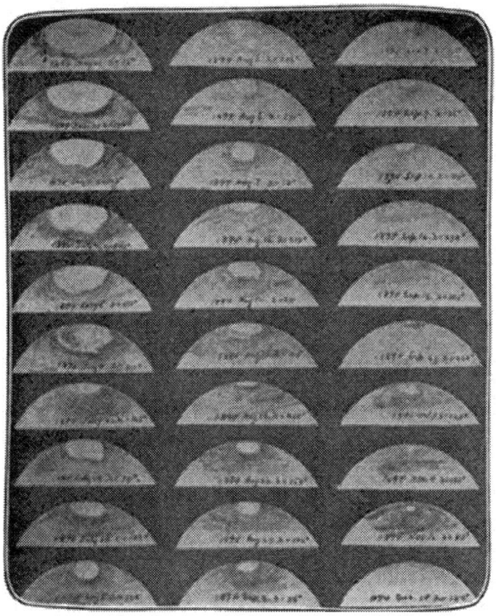

FIG. 97.— The south polar cap of Mars in 1894.— BARNARD.

Sir W. Herschel, who studied these appearances a century ago, compared them with the snow fields which every winter spread out from the region around the terrestrial pole, and in the summer melt and shrink, although with us they do not entirely disappear. This explanation of the polar caps of Mars has been generally accepted among astronomers, and from it we may draw one interesting conclusion: the temperature upon Mars between summer and winter oscillates above and below the freezing point of water, as it does in the temperate zones of the earth. But this conclusion plunges us into a serious difficulty. The temperature of the earth is made by the sun, and at the distance of Mars from the sun the heating effect of the latter is reduced to less than half what it is at the earth, so that, if Mars is to be kept at the same temperature as the earth, there must be some peculiar means for storing the solar heat and using it more economically than is done here. Possibly there is some such mechanism, although no one has yet found it, and some astronomers are very confident that it does not exist, and assert that the comparison of the polar caps with snow fields is misleading, and that the temperature upon Mars must be at least 100°, and perhaps 200° or more, below zero.

153. ***Atmosphere and climate.***— In this connection one feature of Mars is of importance. The markings upon its surface are always visible when turned toward the earth, thus showing that the atmosphere contains no such amount of cloud as does our own, but on the whole is decidedly clear and sunny, and presumably much less dense than ours. We have seen in comparing the earth and the moon how important is the service which the earth's atmosphere renders in storing the sun's heat and checking those great vicissitudes of temperature to which the moon is subject; and with this in mind we must regard the smaller density and cloudless character of the atmosphere of Mars as unfavorable to the maintenance there of a temperature like that of the earth. Indeed, this cloudlessness must mean one of two things: either the temperature is so low that vapors can not exist in any considerable quantity, or the surface of Mars is so dry that there is little water or other liquid to be evaporated. The latter alternative is adopted by those astronomers who look upon the polar caps as true snow fields, which serve as the chief reservoir of the planet's water supply, and who find in Fig. 98 evidence that as the snow melts and the water flows away over the flat, dry surface of the planet, vegetation springs up, as shown by the dark markings on the disk, and gradually dies out with the advancing season. Note that in the first of these pictures the season upon Mars corresponds to the end of May with us, and in the last picture to the beginning of August, a period during which in much of our western country the luxuriant vegetation of spring is burned out by the scorching sun. From this point of view the permanent dark spots are the low-lying parts of the planet's surface, in which at all times there is a sufficient accumulation of water to support vegetable life.

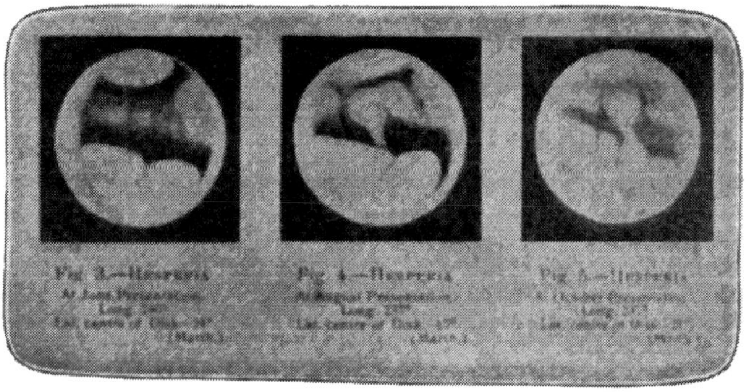

*FIG. 98.— **The same face of Mars at three different seasons.**— LOWELL.*

154. ***The canals.***— In Fig. 98 the lower part of the disk of Mars shows certain faint dark lines which are generally called canals, and in Plate III there

is given a map of Mars showing many of these canals running in narrow, dusky streaks across the face of the planet according to a pattern almost as geometrical as that of a spider's web. This must not be taken for a picture of the planet's appearance in a telescope. No man ever saw Mars look like this, but the map is useful as a plain representation of things dimly seen. Some of the regions of this map are marked Mare (sea), in accordance with the older view which regarded the darker parts of the planet– and of the moon– as bodies of water, but this is now known to be an error in both cases. The curved surface of a planet can not be accurately reproduced upon the flat surface of paper, but is always more or less distorted by the various methods of "projecting" it which are in use. Compare the map of Mars in Plate III with Fig. 99, in which the projection represents very well the equatorial parts of the planet, but enormously exaggerates the region around the poles.

It is a remarkable feature of the canals that they all begin and end in one of these dark parts of the planet's surface; they show no loose ends lying on the bright parts of the planet. Another even more remarkable feature is that while the larger canals are permanent features of the planet's surface, they at times appear "doubled"– i. e., in place of one canal two parallel ones side by side, lasting for a time and then giving place again to a single canal.

It is exceedingly difficult to frame any reasonable explanation of these canals and the varied appearances which they present. The source of the wild speculations about Mars, to which reference is made above, is to be found in the suggestion frequently made, half in jest and half in earnest, that the canals are artificial water courses constructed upon a scale vastly exceeding any public works upon the earth, and testifying to the presence in Mars of an advanced civilization. The distinguished Italian astronomer, Schiaparelli, who has studied these formations longer than any one else, seems inclined to regard them as water courses lined on either side by vegetation, which flourishes as far back from the central channel as water can be supplied from it– a plausible enough explanation if the fundamental difficulty about temperature can be overcome.

155. ***Satellites.***– In 1877, one of the times of near approach, Professor Hall, of Washington, discovered two tiny satellites revolving about Mars in orbits so small that the nearer one, Phobos, presents the remarkable anomaly of completing the circuit of its orbit in less time than the planet takes for a rotation about its axis. This satellite, in fact, makes three revolutions in its orbit while the planet turns once upon its axis, and it therefore rises in the west and sets in the east, as seen from Mars, going from one horizon to the other in a little less than 6 hours.

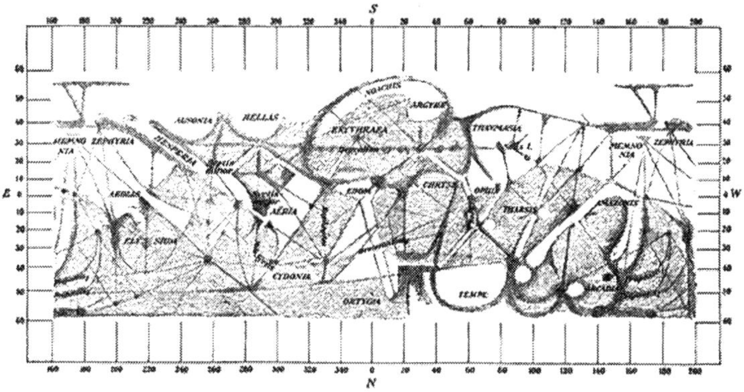

FIG. 99.— A chart of Mars, 1898-'99.— CERULLI.

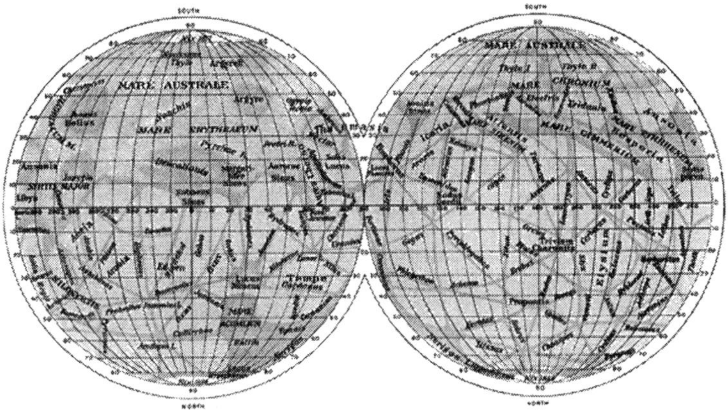

Plate Iii. Map Of Mars (After Schiaparelli)

The other satellite, Deimos, takes a few hours more than a day to make the circuit of its orbit, but the difference is so small that it remains continuously above the horizon of any given place upon Mars for more than 60 hours at a time, and during this period runs twice through its complete set of phases—new, first quarter, full, etc. In ordinary telescopes these satellites can be seen only under especially favorable circumstances, and are far too small to permit of any direct measurement of their size. The amount of light which they reflect has been compared with that of Mars and found to be as much inferior to it as is Polaris to two full moons, and, judging from this comparison, their diameters can not much exceed a half dozen miles, unless their albedo is far less than that of Mars, which does not seem probable.

The Asteroids

156. ***Minor planets.***— These may be dismissed with few words. There are about 500 of them known, all discovered since the beginning of the nineteenth century, and new ones are still found every year. No one pretends to remember the names which have been assigned them, and they are commonly represented by a number inclosed in a circle, showing the order in which they were discovered— e. g., ① = Ceres, [circle 433] = Eros, etc. For the most part they are little more than chips, world fragments, adrift in space, and naturally it was the larger and brighter of them that were first discovered. The size of the first four of them— Ceres, Pallas, Juno, and Vesta— compared with the size of the moon, according to Professor Barnard, is shown in Fig. 100. The great majority of them must be much smaller than the smallest of these, perhaps not more than a score of miles in diameter.

A few of the asteroids present problems of special interest, such as Eros, on account of its close approach to the earth; Polyhymnia, whose very eccentric orbit makes it a valuable means for determining the mass of Jupiter, etc.; but these are special cases and the average asteroid now receives scant attention, although half a century ago, when only a few of them were known, they were regarded with much interest, and the discovery of a new one was an event of some consequence.

It was then a favorite speculation that they were in fact fragments of an ill-fated planet which once filled the gap between the orbits of Mars and Jupiter, but which, by some mischance, had been blown into pieces. This is now known to be well-nigh impossible, for every fragment which after the explosion moved in an elliptical orbit, as all the asteroids do move, would be brought back once in every revolution to the place of the explosion, and all the asteroid orbits must therefore intersect at this place. But there is no such common point of intersection.

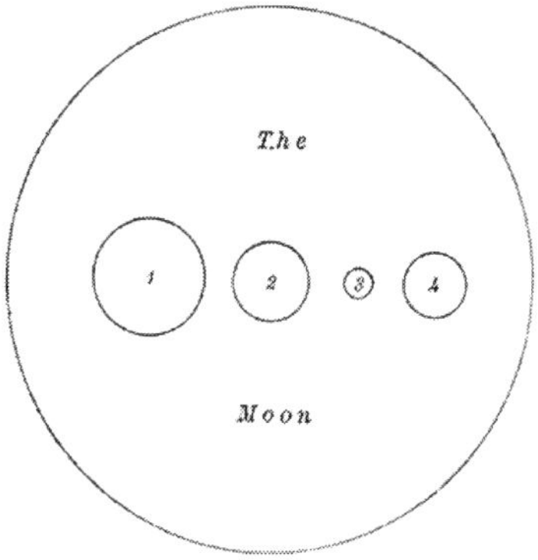

FIG. 100.— The size of the first four asteroids.— BARNARD.

157. ***Life on the planets.*** – There is a belief firmly grounded in the popular mind, and not without its advocates among professional astronomers, that the planets are inhabited by living and intelligent beings, and it seems proper at the close of this chapter to inquire briefly how far the facts and principles here developed are consistent with this belief, and what support, if any, they lend to it.

At the outset we must observe that the word life is an elastic term, hard to define in any satisfactory way, and yet standing for something which we know here upon the earth. It is this idea, our familiar though crude knowledge of life, which lies at the root of the matter. Life, if it exists in another planet, must be in its essential character like life upon the earth, and must at least possess those features which are common to all forms of terrestrial life. It is an abuse of language to say that life in Mars may be utterly unlike life in the earth; if it is absolutely unlike, it is not life, whatever else it may be. Now, every form of life found upon the earth has for its physical basis a certain chemical compound, called protoplasm, which can exist and perpetuate itself only within a narrow range of temperature, roughly speaking, between 0° and 100° centigrade, although these limits can be considerably overstepped for short periods of time. Moreover, this protoplasm can be active only in the presence of water, or water vapor, and we may therefore establish as the necessary conditions for the continued existence and reproduction of life in any place that its temperature must not be permanently above 100° or below 0°, C., and water must be present in that place in some form.

With these conditions before us it is plain that life can not exist in the sun on account of its high temperature. It is conceivable that active and intelligent beings, salamanders, might exist there, but they could not properly be said to live. In Jupiter and Saturn the same condition of high temperature prevails, and probably also in Uranus and Neptune, so that it seems highly improbable that any of these planets should be the home of life.

Of the inner planets, Mercury and the moon seem destitute of any considerable atmospheres, and are therefore lacking in the supply of water necessary for life, and the same is almost certainly true of all the asteroids. There remain Venus, Mars, and the satellites of the outer planets, which latter, however, we must drop from consideration as being too little known. On Venus there is an atmosphere probably containing vapor of water, and it is well within the range of possibility that liquid water should exist upon the surface of this planet and that its temperature should fall within the prescribed limits. It would, however, be straining our actual knowledge to affirm that such is the case, or to insist that if such were the case, life would necessarily exist upon the planet.

On Mars we encounter the fundamental difficulty of temperature already noted in § 152. If in some unknown way the temperature is maintained sufficiently high for the polar caps to be real snow, thawing and forming again with the progress of the seasons, the necessary conditions of life would seem to be fulfilled here and life if once introduced upon the planet might abide and flourish. But of positive proof that such is the case we have none.

On the whole, our survey lends little encouragement to the belief in planetary life, for aside from the earth, of all the hundreds of bodies in the solar system, not one is found in which the necessary conditions of life are certainly fulfilled, and only two exist in which there is a reasonable probability that these conditions may be satisfied.

CHAPTER XII.
COMETS AND METEORS

158. ***Visitors in the solar system.***— All of the objects— sun, moon, planets, stars— which we have thus far had to consider, are permanent citizens of the sky, and we have no reason to suppose that their present appearance differs appreciably from what it was 1,000 years or 10,000 years ago. But there is another class of objects— comets, meteors— which appear unexpectedly, are visible for a time, and then vanish and are seen no more. On account of this temporary character the astronomers of ancient and mediæval times for the most part refused to regard them as celestial bodies but classed them along with clouds, fogs, Jack-o'-lanterns, and fireflies, as exhalations from the swamps or the volcano; admitting them to be indeed important as harbingers of evil to mankind, but having no especial significance for the astronomer.

The comet of 1618 A. D. inspired the lines—

"Eight things there be a Comet brings,When it on high doth horrid range:Wind, Famine, Plague, and Death to Kings,War, Earthquakes, Floods, and Direful Change,"

which, according to White (History of the Doctrine of Comets), were to be taught in all seriousness to peasants and school children.

It was by slow degrees, and only after direct measurements of parallax had shown some of them to be more distant than the moon, that the tide of old opinion was turned and comets were transferred from the sublunary to the celestial sphere, and in more recent times meteors also have been recognized as coming to us from outside the earth. A meteor, or shooting star as it is often called, is one of the commonest of phenomena, and one can hardly watch the sky for an hour on any clear and moonless night without seeing several of those quick flashes of light which look as if some star had suddenly left its place, dashed swiftly across a portion of the sky and then vanished. It is this misleading appearance that probably is responsible for the name shooting star.

159. ***Comets.***— Comets are less common and much longer-lived than meteors, lasting usually for several weeks, and may be visible night after night for many months, but never for many years, at a time. During the last decade there is no year in which less than three comets have appeared, and 1898 is distinguished by the discovery of ten of these bodies, the largest number ever found in one year. On the average, we may expect a new comet to be found about once in every ten weeks, but for the most part they are small affairs, visible only in the telescope, and a fine large one, like Donati's comet of 1858 (Fig. 101), or the Great Comet of September, 1882, which was visible in broad daylight close beside the sun, is a rare spectacle, and as striking and impressive as it is rare.

FIG. 101.— Donati's comet.— BOND.

159. ***Comets.***— Comets are less common and much longer-lived than meteors, lasting usually for several weeks, and may be visible night after night for many months, but never for many years, at a time. During the last decade there is no year in which less than three comets have appeared, and 1898 is distinguished by the discovery of ten of these bodies, the largest number ever found in one year. On the average, we may expect a new comet to be found about once in every ten weeks, but for the most part they are small affairs, visible only in the telescope, and a fine large one, like Donati's comet of 1858 (Fig. 101), or the Great Comet of September, 1882, which was visible in broad daylight close beside the sun, is a rare spectacle, and as striking and impressive as it is rare.

Note in Fig. 102 the great variety of aspect presented by some of the more famous comets, which are here represented upon a very small scale.

Fig. 103 is from a photograph of one of the faint comets of the year 1893, which appears here as a rather feeble streak of light amid the stars which are scattered over the background of the picture. An apparently detached portion of this comet is shown at the extreme left of the picture, looking almost like another independent comet. The clean, straight line running diagonally across the picture is the flash of a bright meteor that chanced to pass within the range of the camera while the comet was being photographed.

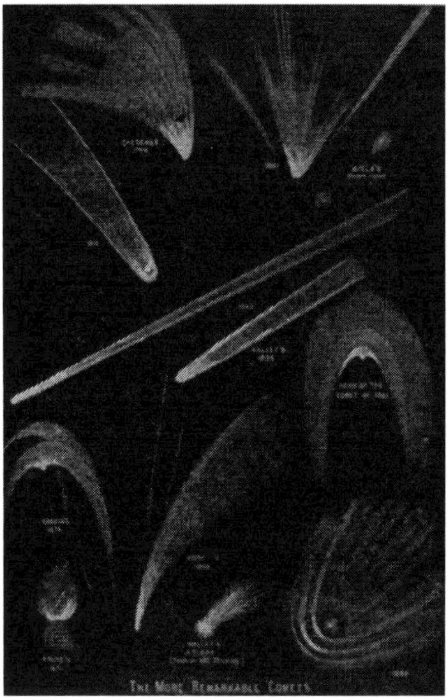

FIG. 102.— Some famous Comets.

FIG. 103.— Brooks's comet, November 13, 1893. BARNARD.

A more striking representation of a moderately bright telescopic comet is contained in Figs. 104 and 105, which present two different views of the same comet, showing a considerable change in its appearance. A striking feature of Fig. 105 is the star images, which are here drawn out into short lines all parallel with each other. During the exposure of 2h. 20m. required to imprint this picture upon the photographic plate, the comet was continually changing its position among the stars on account of its orbital motion, and the plate was therefore moved from time to time, so as to follow the comet and make its image always fall at the same place. Hence the plate was continually shifted relative to the stars whose images, drawn out into lines, show the direction in which the plate was moved– i. e., the direction in which the comet was moving across the sky. The same effect is shown in the other photographs, but less conspicuously than here on account of their shorter exposure times.

These pictures all show that one end of the comet is brighter and apparently more dense than the other, and it is customary to call this bright part the *head* of the comet, while the brushlike appendage that streams away from it is called the comet's *tail*.

160. ***The parts of a comet.***– It is not every comet that has a tail, though all the large ones do, and in Fig. 103 the detached piece of cometary matter at the left of the picture represents very well the appearance of a tailless comet, a rather large but not very bright star of a fuzzy or hairy appearance. The word comet means long-haired or hairy star. Something of this vagueness of outline is found in all comets, whose exact boundaries are hard to define, instead of being sharp and clean-cut like those of a planet or satellite. Often, however, there is found in the head of a comet a much more solid appearing part, like the round white ball at the center of Fig. 106, which is called the nucleus of the comet, and appears to be in some sort the center from which its activities radiate. As shown in Figs. 106 and 107, the nucleus is sometimes surrounded by what are called envelopes, which have the appearance of successive wrappings or halos placed about it, and odd, spurlike projections, called jets, are sometimes found in connection with the envelopes or in place of them. These figures also show what is quite a common characteristic of large comets, a dark streak running down the axis of the tail, showing that the tail is hollow, a mere shell surrounding empty space.

The amount of detail shown in Figs. 106 and 107 is, however, quite exceptional, and the ordinary comet is much more like Fig. 103 or 104. Even a great comet when it first appears is not unlike the detached fragment in Fig. 103, a faint and roundish patch of foggy light which grows through successive stages to its maximum estate, developing a tail, nucleus, envelopes, etc., only to lose them again as it shrinks and finally disappears.

FIG. 104.— Swift's comet, April 17, 1892.— BARNARD.

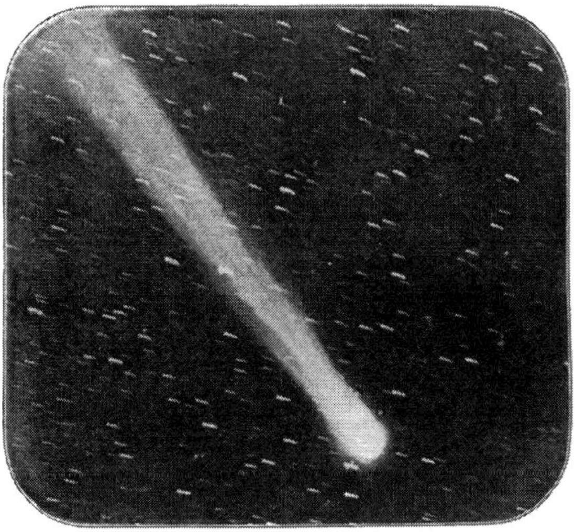

FIG. 105.— Swift's comet, April 24, 1892.— BARNARD.

161. ***The orbits of comets.***— It will be remembered that Newton found, as a theoretical consequence of the law of gravitation, that a body moving under the influence of the sun's attraction might have as its orbit any one of the conic sections, ellipse, parabola, or hyperbola, and among the 400 and more comet orbits which have been determined every one of these orbit forms appears, but curiously enough there is not a hyperbola among them which, if drawn upon paper, could be distinguished by the unaided eye from

a parabola, and the ellipses are all so long and narrow, not one of them being so nearly round as is the most eccentric planet orbit, that astronomers are accustomed to look upon the parabola as being the normal type of comet orbit, and to regard a comet whose motion differs much from a parabola as being abnormal and calling for some special explanation.

FIG. 106.—Head of Coggia's comet, July 13, 1874.—TROUVELOT.

The fact that comet orbits are parabolas, or differ but little from them, explains at once the temporary character and speedy disappearance of these bodies. They are visitors to the solar system and visible for only a short time, because the parabola in which they travel is not a closed curve, and the comet, having passed once along that portion of it near the earth and the sun, moves off along a path which ever thereafter takes it farther and farther away, beyond the limit of visibility. The development of the comet during the time it is visible, the growth and disappearance of tail, nucleus, etc., depend upon its changing distance from the sun, the highest development and most complex structure being presented when it is nearest to the sun.

Fig. 108 shows the path of the Great Comet of 1882 during the period in which it was seen, from September 3, 1882, to May 26, 1883. These dates—IX, 3, and V, 26— are marked in the figure opposite the parts of the orbit in which the comet stood at those times. Similarly, the positions of the earth in its orbit at the beginning of September, October, November, etc., are marked by the Roman numerals IX, X, XI, etc. The line *S V* shows the direction from the sun to the vernal equinox, and *S* Ω is the line along which the plane of the comet's orbit intersects the plane of the earth's orbit— i. e., it is the line of nodes of the comet orbit. Since the comet approached the sun from the south side of the ecliptic, all of its orbit, save the little segment which falls to the left of *S* Ω, lies below (south) of the plane of the earth's orbit, and the part which would be hidden if this plane were opaque is represented by a broken line.

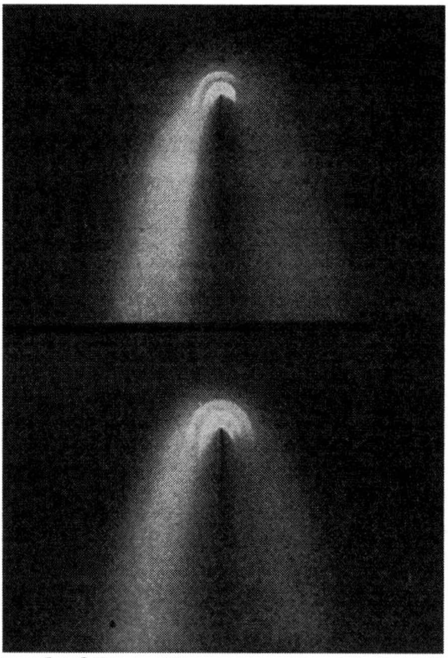

*FIG. 107.— **Head of Donati's comet, September 30, October 2, 1858.**— BOND.*

162. ***Elements of a comet's orbit.***— There is a theorem of geometry to the effect that through any three points not in the same straight line one circle, and only one, can be drawn. Corresponding to this there is a theorem of celestial mechanics, that through any three positions of a comet one conic section, and only one, can be passed along which the comet can move in accordance with the law of gravitation. This conic section is, of course, its orbit, and at the discovery of a comet astronomers always hasten to observe its position in the sky on different nights in order to obtain the three positions

(right ascensions and declinations) necessary for determining the particular orbit in which it moves. The circle, to which reference was made above, is completely ascertained and defined when we know its radius and the position of its center. A parabola is not so simply defined, and five numbers, called the *elements* of its orbit, are required to fix accurately a comet's path around the sun. Two of these relate to the position of the line of nodes and the angle which the orbit plane makes with the plane of the ecliptic; a third fixes the direction of the axis of the orbit in its plane, and the remaining two, which are of more interest to us, are the date at which the comet makes its nearest approach to the sun (*perihelion passage*) and its distance from the sun at that date (*perihelion distance*). The date, September 17th, placed near the center of Fig. 108, is the former of these elements, while the latter, which is too small to be accurately measured here, may be found from Fig. 109 to be 0.82 of the sun's diameter, or, in terms of the earth's distance from the sun, 0.008.

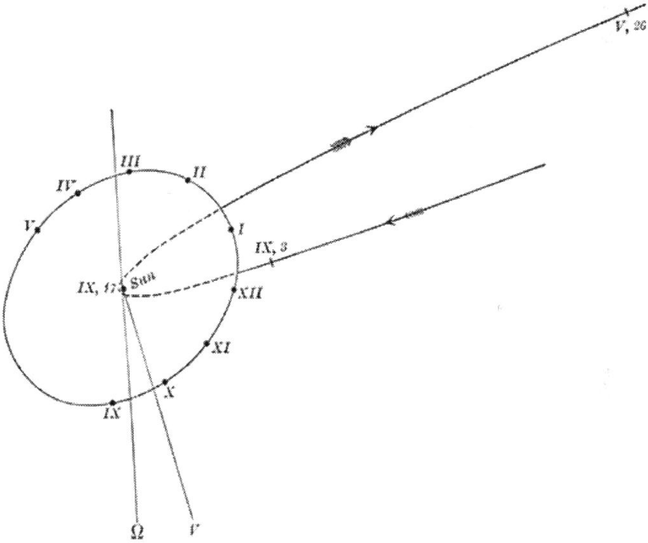

FIG. 108.— Orbits of the earth and the Great Comet of 1882.

Fig. 109 shows on a large scale the shape of that part of the orbit near the sun and gives the successive positions of the comet, at intervals of 2/10 of a day, on September 16th and 17th, showing that in less than 10 hours— 17.0 to 17.4— the comet swung around the sun through an angle of more than 240°. When at its perihelion it was moving with a velocity of 300 miles per second! This very unusual velocity was due to the comet's extraordinarily close approach to the sun. The earth's velocity in its orbit is only 19 miles per second, and the velocity of any comet at any distance from the sun, provided its orbit is a parabola, may be found by dividing this number by the square root of half the comet's distance– e. g., 300 miles per second equals $19 \div \sqrt{0.004}$.

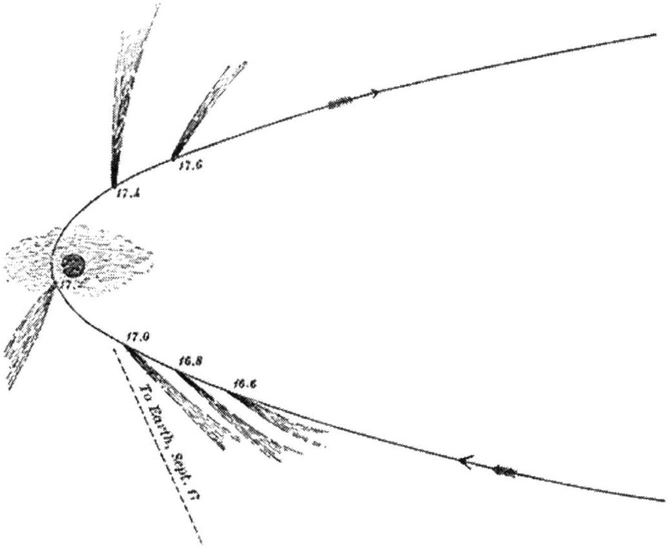

FIG. 109.— Motion of the Great Comet of 1883 in passing around the sun.

Most of the visible comets have their perihelion distances included between 1/3 and 4/3 of the earth's distance from the sun, but occasionally one is found, like the second comet of 1885, whose nearest approach to the sun lies far outside the earth's orbit, in this case half-way out to the orbit of Jupiter; but such a comet must be a very large one in order to be seen at all from the earth. There is, however, some reason for believing that the number of comets which move around the sun without ever coming inside the orbit of Jupiter, or even that of Saturn, is much larger than the number of those which come close enough to be discovered from the earth. In any case we are reminded of Kepler's saying, that comets in the sky are as plentiful as fishes in the sea, which seems to be very little exaggerated when we consider that, according to Kleiber, out of all the comets which enter the solar system probably not more than 2 or 3 per cent are ever discovered.

FIG. 110.— The Great Comet of 1843.

163. ***Dimensions of comets.***— The comet whose orbit is shown in Figs. 108 and 109 is the finest and largest that has appeared in recent years. Its tail, which at its maximum extent would have more than bridged the space between sun and earth (100,000,000 miles), is made very much too short in Fig. 109, but when at its best was probably not inferior to that of the Great Comet of 1843, shown in Fig. 110. As we shall see later, there is a peculiar and special relationship between these two comets.

The head of the comet of 1882 was not especially large— about twice the diameter of the ball of Saturn— but its nucleus, according to an estimate made by Dr. Elkin when it was very near perihelion, was as large as the moon. The head of the comet shown in Fig. 107 was too large to be put in the space between the earth and the moon, and the Great Comet of 1811 had a head considerably larger than the sun itself. From these colossal sizes down to the smallest shred just visible in the telescope, comets of all dimensions may be found, but the smaller the comet the less the chance of its being discovered, and a comet as small as the earth would probably go unobserved unless it approached very close to us.

164. ***The mass of a comet.***— There is no known case in which the mass of a comet has ever been measured, yet nothing about them is more sure than that they are bodies with mass which is attracted by the sun and the planets, and which in its turn attracts both sun and planets and produces perturbations in their motion. These perturbations are, however, too small to be measured,

although the corresponding perturbations in the comet's motion are sometimes enormous, and since these mutual perturbations are proportional to the masses of comet and planet, we are forced to say that, by comparison with even such small bodies as the moon or Mercury, the mass of a comet is utterly insignificant, certainly not as great as a ten-thousandth part of the mass of the earth. In the case of the Great Comet of 1882, if we leave its hundred million miles of tail out of account and suppose the entire mass condensed into its head, we find by a little computation that the average density of the head under these circumstances must have been less than 1/1500 of the density of air. In ordinary laboratory practice this would be called a pretty good vacuum. A striking observation made on September 17, 1882, goes to confirm the very small density of this comet. It is shown in Fig. 109 that early on that day the comet crossed the line joining earth and sun, and therefore passed in transit over the sun's disk. Two observers at the Cape of Good Hope saw the comet approach the sun, and followed it with their telescopes until the nucleus actually reached the edge of the sun and disappeared, behind it as they supposed, for no trace of the comet, not even its nucleus, could be seen against the sun, although it was carefully looked for. Now, the figure shows that the comet passed between the earth and sun, and its densest parts were therefore too attenuated to cut off any perceptible fraction of the sun's rays. In other cases stars have been seen through the head of a comet, shining apparently with undimmed luster, although in some cases they seem to have been slightly refracted out of their true positions.

165. ***Meteors.***— Before proceeding further with the study of comets it is well to turn aside and consider their humbler relatives, the shooting stars. On some clear evening, when the moon is absent from the sky, watch the heavens for an hour and count the meteors visible during that time. Note their paths, the part of the sky where they appear and where they disappear, their brightness, and whether they all move with equal swiftness. Out of such simple observations with the unaided eye there has grown a large and important branch of astronomical science, some parts of which we shall briefly summarize here.

A particular meteor is a local phenomenon seen over only a small part of the earth's surface, although occasionally a very big and bright one may travel and be visible over a considerable territory. Such a one in December, 1876, swept over the United States from Kansas to Pennsylvania, and was seen from eleven different States. But the ordinary shooting star is much less conspicuous, and, as we know from simultaneous observations made at neighboring places, it makes its appearance at a height of some 75 miles above the earth's surface, occupies something like a second in moving over its path, and then disappears at a height of about 50 miles or more, although occasionally a big one comes down to the very surface of the earth with force sufficient to bury itself in the ground, from which it may be dug up, handled, weighed, and turned over to the chemist to be analyzed. The pieces thus found show that the big meteors, at least, are masses of stone or mineral; iron is quite commonly found in them, as are a considerable number of other terrestrial substances combined in rather peculiar ways. But no chemical element not found on the earth has ever been discovered in a meteor.

166. ***Nature of meteors.***— The swiftness with which the meteors sweep down shows that they must come from outside the earth, for even half their velocity, if given to them by some terrestrial volcano or other explosive agent, would send them completely away from the earth never to return. We must therefore look upon them as so many projectiles, bullets, fired against the earth from some outside source and arrested in their motion by the earth's atmosphere, which serves as a cushion to protect the ground from the bombardment which would otherwise prove in the highest degree dangerous to both property and life. The speed of the meteor is checked by the resistance which the atmosphere offers to its motion, and the energy represented by that speed is transformed into heat, which in less than a second raises the meteor and the surrounding air to incandescence, melts the meteor either wholly or in part, and usually destroys its identity, leaving only an impalpable dust, which cools off as it settles slowly through the lower atmosphere to the ground. The heating effect of the air's resistance is proportional to the square of the meteor's velocity, and even at such a moderate speed as 1 mile per second the effect upon the meteor is the same as if it stood still in a bath of red-hot air. Now, the actual velocity of meteors through the air is often 30 or 40 times as great as this, and the corresponding effect of the air in raising its temperature is more than 1,000 times that of red heat. Small wonder that the meteor is brought to lively incandescence and consumed even in a fraction of a second.

167. ***The number of meteors.***— A single observer may expect to see in the evening hours about one meteor every 10 minutes on the average, although, of course, in this respect much irregularity may occur. Later in the night they become more frequent, and after 2 A. M. there are about three times as many to be seen as in the evening hours. But no one person can keep a watch upon the whole sky, high and low, in front and behind, and experience shows that by increasing the number of observers and assigning to each a particular part of the sky, the total number of meteors counted may be increased about five-fold. So, too, the observers at any one place can keep an effective watch upon only those meteors which come into the earth's atmosphere within some moderate distance of their station, say 50 or 100 miles, and to watch every part of that atmosphere would require a large number of stations, estimated at something more than 10,000, scattered systematically over the whole face of the earth. If we piece together the several numbers above considered, taking 14 as a fair average of the hourly number

of meteors to be seen by a single observer at all hours of the night, we shall find for the total number of meteors encountered by the earth in 24 hours, $14 \times 5 \times 10{,}000 \times 24 = 16{,}800{,}000$. Without laying too much stress upon this particular number, we may fairly say that the meteors picked up by the earth every day are to be reckoned by millions, and since they come at all seasons of the year, we shall have to admit that the region through which the earth moves, instead of being empty space, is really a dust cloud, each individual particle of dust being a prospective meteor.

On the average these individual particles are very small and very far apart; a cloud of silver dimes each about 250 miles from its nearest neighbor is perhaps a fair representation of their average mass and distance from each other, but, of course, great variations are to be expected both in the size and in the frequency of the particles. There must be great numbers of them that are too small to make shooting stars visible to the naked eye, and such are occasionally seen darting by chance across the field of view of a telescope.

168. ***The zodiacal light*** is an effect probably due to the reflection of sunlight from the myriads of these tiny meteors which occupy the space inside the earth's orbit. It is a faint and diffuse stream of light, something like the Milky Way, which may be seen in the early evening or morning stretching up from the sunrise or sunset point of the horizon along the ecliptic and following its course for many degrees, possibly around the entire circumference of the sky. It may be seen at any season of the year, although it shows to the best advantage in spring evenings and autumn mornings. Look for it.

169. ***Great meteors.***— But there are other meteors, veritable fireballs in appearance, far more conspicuous and imposing than the ordinary shooting star. Such a one exploded over the city of Madrid, Spain, on the morning of February 10, 1896, giving in broad sunlight "a brilliant flash which was followed ninety seconds later by a succession of terrific noises like the discharge of a battery of artillery." Fig. 111 shows a large meteor which was seen in California in the early evening of July 27, 1894, and which left behind it a luminous trail or cloud visible for more than half an hour.

Not infrequently large meteors are found traveling together, two or three or more in company, making their appearance simultaneously as did the California meteor of October 22, 1896, which is described as triple, the trio following one another like a train of cars, and Arago cites an instance, from the year 1830, where within a short space of time some forty brilliant meteors crossed the sky, all moving in the same direction with a whistling noise and displaying in their flight all the colors of the rainbow.

The mass of great meteors such as these must be measured in hundreds if not thousands of pounds, and stories are current, although not very well authenticated, of even larger ones, many tons in weight, having been found partially buried in the ground. Of meteors which have been actually seen to fall from the sky, the largest single fragment recovered weighs about 500 pounds, but it is only a fragment of the original meteor, which must have been much more massive before it was broken up by collision with the atmosphere.

FIG. 111.— The California meteor of July 27, 1894.

170. ***The velocity of meteors.***— Every meteor, big or little, is subject to the law of gravitation, and before it encounters the earth must be moving in some kind of orbit having the sun at its focus, the particular species of orbit— ellipse, parabola, hyperbola— depending upon the velocity and direction of its motion. Now, the direction in which a meteor is moving can be determined without serious difficulty from observations of its apparent path across the sky made by two or more observers, but the velocity can not be so readily found, since the meteors go too fast for any ordinary process of timing. But by photographing one of them two or three times on the same plate, with an interval of only a tenth of a second between exposures, Dr. Elkin has succeeded in showing, in a few cases, that their velocities varied from 20 to

25 miles per second, and must have been considerably greater than this before the meteors encountered the earth's atmosphere. This is a greater velocity than that of the earth in its orbit, 19 miles per second, as might have been anticipated, since the mere fact that meteors can be seen at all in the evening hours shows that some of them at least must travel considerably faster than the earth, for, counting in the direction of the earth's motion, the region of sunset and evening is always on the rear side of the earth, and meteors in order to strike this region must overtake it by their swifter motion. We have here, in fact, the reason why meteors are especially abundant in the morning hours; at this time the observer is on the front side of the earth which catches swift and slow meteors alike, while the rear is pelted only by the swifter ones which follow it.

A comparison of the relative number of morning and evening meteors makes it probable that the average meteor moves, relative to the sun, with a velocity of about 26 miles per second, which is very approximately the average velocity of comets when they are at the earth's distance from the sun. Astronomers, therefore, consider meteors as well as comets to have the parabola and the elongated ellipse as their characteristic orbits.

171. ***Meteor showers— The radiant.***— There is evident among meteors a distinct tendency for individuals, to the number of hundreds or even hundreds of millions, to travel together in flocks or swarms, all going the same way in orbits almost exactly alike. This gregarious tendency is made manifest not only by the fact that from time to time there are unusually abundant meteoric displays, but also by a striking peculiarity of their behavior at such times. The meteors all seem to come from a particular part of the heavens, as if here were a hole in the sky through which they were introduced, and from which they flow away in every direction, even those which do not visibly start from this place having paths among the stars which, if prolonged backward, would pass through it. The cause of this appearance may be understood from Fig. 112, which represents a group of meteors moving together along parallel paths toward an observer at *D*. Traveling unseen above the earth until they encounter the upper strata of its atmosphere, they here become incandescent and speed on in parallel paths, *1*, *2*, *3*, *4*, *5*, *6*, which, as seen by the observer, are projected back against the sky into luminous streaks that, as is shown by the arrowheads, *b*, *c*, *d*, all seem to radiate from the point *a*— i. e., from the point in the sky whose direction from the observer is parallel to the paths of the meteors.

Such a display is called a meteor shower, and the point *a* is called its radiant. Note how those meteors which appear near the radiant all have short paths, while those remote from it in the sky have longer ones. Query: As the night wears on and the stars shift toward the west, will the radiant share in their motion or will it be left behind? Would the luminous part of the path of any of these meteors pass across the radiant from one side to the other?

Is such a crossing of the radiant possible under any circumstances?

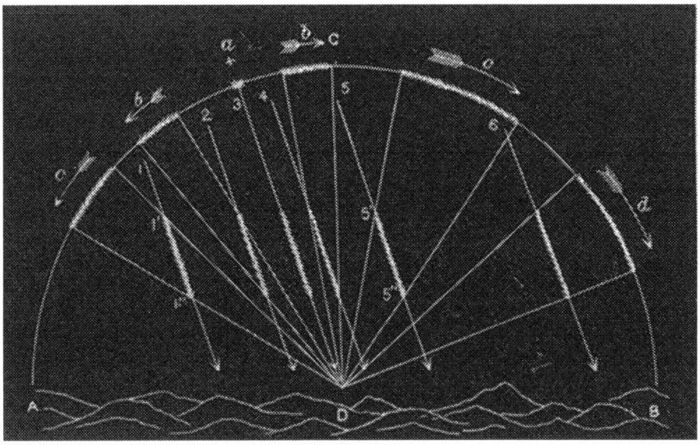

FIG. 112.— Explanation of the radiant of a meteoric shower.— DENNING.

Fig. 113 shows how the meteor paths are grouped around the radiant of a strongly marked shower. Select from it the meteors which do not belong to this shower.

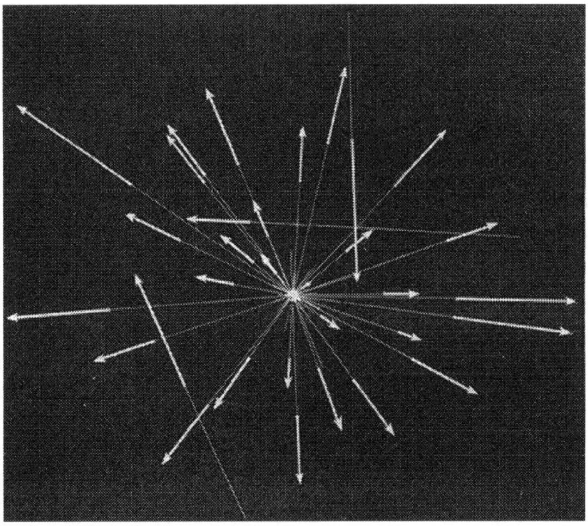

FIG. 113.— The radiant of a meteoric shower, showing also the paths of three meteors which do not belong to this shower.— DENNING.

Many hundreds of these radiants have been observed in the sky, each of which represents an orbit along which a group of meteors moves, and the relation of one of these orbits to that of the earth is shown in Fig. 114. The orbit of the meteors is an ellipse extending out beyond the orbit of Uranus, but so eccentric that a part of it comes inside the orbit of the earth, and the figure shows only that part of it which lies nearest the sun. The Roman numerals which are placed along the earth's orbit show the position of the earth at the beginning of the tenth month, eleventh month, etc. The meteors flow along their orbit in a long procession, whose direction of motion is indicated by the arrow heads, and the earth, coming in the opposite direction, plunges into this stream and receives the meteor shower when it reaches the intersection of the two orbits. The long arrow at the left of the figure represents the direction of motion of another meteor shower which encounters the earth at this point.

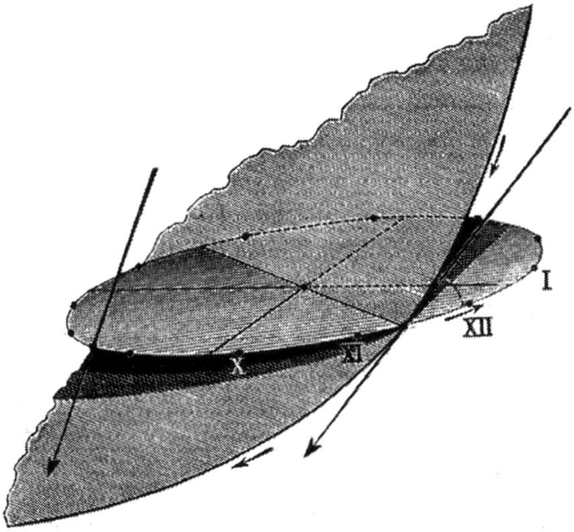

*FIG. 114.— **The orbits of the earth and the November meteors.***

Can you determine from the figure answers to the following questions? On what day of the year will the earth meet each of these showers? Will the radiant points of the showers lie above or below the plane of the earth's orbit? Will these meteors strike the front or the rear of the earth? Can they be seen in the evening hours?

From many of the radiants year after year, upon the same day or week in each year, there comes a swarm of shooting stars, showing that there must be a continuous procession of meteors moving along this orbit, so that some are always ready to strike the earth whenever it reaches the intersection of its orbit with theirs. Such is the explanation of the shower which appears each year in the first half of August, and whose meteors are sometimes called Perseids, because their radiant lies in the constellation Perseus, and a similar explanation holds for all the star showers which are repeated year after year.

172. ***The Leonids.***— There is, however, a kind of star shower, of which the Leonids (radiant in Leo) is the most conspicuous type, in which the shower, although repeated from year to year, is much more striking in some

years than in others. Thus, to quote from the historian: "In 1833 the shower was well observed along the whole eastern coast of North America from the Gulf of Mexico to Halifax. The meteors were most numerous at about 5 A. M. on November 13th, and the rising sun could not blot out all traces of the phenomena, for large meteors were seen now and then in full daylight. Within the scope that the eye could contain, more than twenty could be seen at a time shooting in every direction. Not a cloud obscured the broad expanse, and millions of meteors sped their way across in every point of the compass. Their coruscations were bright, gleaming, and incessant, and they fell thick as the flakes in the early snows of December." But, so far as is known, none of them reached the ground. An illiterate man on the following day remarked: "The stars continued to fall until none were left. I am anxious to see how the heavens will appear this evening, for I believe we shall see no more stars."

An eyewitness in the Southern States thus describes the effect of this shower upon the plantation negroes: "Upward of a hundred lay prostrate upon the ground, some speechless and some with the bitterest cries, but with their hands upraised, imploring God to save the world and them. The scene was truly awful, for never did rain fall much thicker than the meteors fell toward the earth– east, west, north, and south it was the same." In the preceding year a similar but feebler shower from the same radiant created much alarm in France, and through the old historic records its repetitions may be traced back at intervals of 33 or 34 years, although with many interruptions, to October 12, 902, O. S., when "an immense number of falling stars were seen to spread themselves over the face of the sky like rain."

Such a star shower differs from the one repeated every year chiefly in the fact that its meteors, instead of being drawn out into a long procession, are mainly clustered in a single flock which may be long enough to require two or three or four years to pass a given point of its orbit, but which is far from extending entirely around it, so that meteors from this source are abundant only in those years in which the flock is at or near the intersection of its orbit with that of the earth. The fact that the Leonid shower is repeated at intervals of 33 or 34 years (it appeared in 1799, 1832-'33, 1866-'67) shows that this is the "periodic time" in its orbit, which latter must of course be an ellipse, and presumably a long and narrow one. It is this orbit which is shown in Fig. 114, and the student should note in this figure that if the meteor stream at the point where it cuts through the plane of the earth's orbit were either nearer to or farther from the sun than is the earth there could be no shower; the earth and the meteors would pass by without a collision. Now, the meteors in their motion are subject to perturbations, particularly by the large planets Jupiter, Saturn, and Uranus, which slightly change the meteor orbit, and it seems certain that the changes thus produced will sometimes thrust the swarm inside or outside the orbit of the earth, and thus cause a failure of the shower at times when it is expected. The meteors were due at the crossing of the orbits in November, 1899 and 1900, and, although a few were then seen, the shower was far from being a brilliant one, and its failure was doubtless caused by the outer planets, which switched the meteors aside from the path in which they had been moving for a century. Whether they will be again switched back so as to produce future showers is at the present time uncertain.

173. ***Capture of the Leonids.***— But a far more striking effect of perturbations is to be found in Fig. 115, which shows the relation of the Leonid orbit to those of the principal planets, and illustrates a curious chapter in the history of the meteor swarm that has been worked out by mathematical analysis, and is probably a pretty good account of what actually befell them. Early in the second century of the Christian era this flock of meteors came down toward the sun from outer space, moving along a parabolic orbit which would have carried it just inside the orbit of Jupiter, and then have sent it off to return no more. But such was not to be its fate. As it approached the orbit of Uranus, in the year 126 A. D., that planet chanced to be very near at hand and perturbed the motion of the meteors to such an extent that the character of their orbit was completely changed into the ellipse shown in the figure, and in this new orbit they have moved from that time to this, permanent instead of transient members of the solar system. The perturbations, however, did not end with the year in which the meteors were captured and annexed to the solar system, but ever since that time Jupiter, Saturn, and Uranus have been pulling together upon the orbit, and have gradually turned it around into its present position as shown in the figure, and it is chiefly this shifting of the orbit's position in the thousand years that have elapsed since 902 A. D. that makes the meteor shower now come in November instead of in October as it did then.

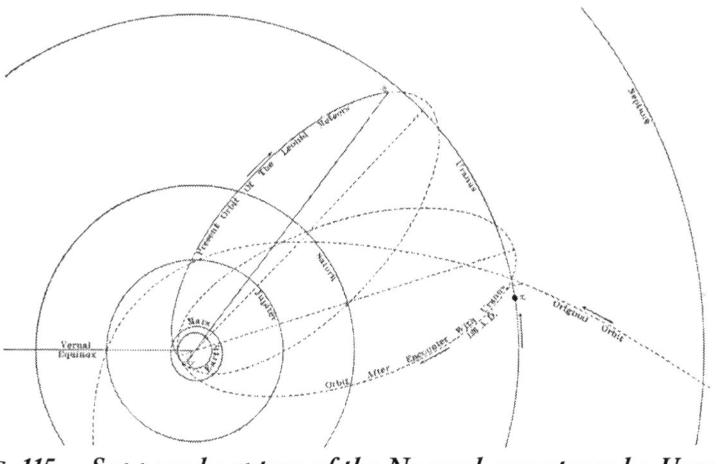

FIG. 115.— Supposed capture of the November meteors by Uranus.

174. ***Breaking up a meteor swarm.***— How closely packed together these meteors were at the time of their annexation to the solar system is unknown, but it is certain that ever since that time the sun has been exerting upon them a tidal influence tending to break up the swarm and distribute its particles around the orbit, as the Perseids are distributed, and, given sufficient time, it will accomplish this, but up to the present the work is only partly done. A certain number of the meteors have gained so much over the slower moving ones as to have made an extra circuit of the orbit and overtaken the rear of the procession, so that there is a thin stream of them extending entirely around the orbit and furnishing in every November a Leonid shower; but by far the larger part of the meteors still cling together, although drawn out into a stream or ribbon, which, though very thin, is so long that it takes some three years to pass through the perihelion of its orbit. It is only when the earth plunges through this ribbon, as it should in 1899, 1900, 1901, that brilliant Leonid showers can be expected.

175. ***Relation of comets and meteors.***— It appears from the foregoing that meteors and comets move in similar orbits, and we have now to push the analogy a little further and note that in some instances at least they move in identically the same orbit, or at least in orbits so like that an appreciable difference between them is hardly to be found. Thus a comet which was discovered and observed early in the year 1866, moves in the same orbit with the Leonid meteors, passing its perihelion about ten months ahead of the main body of the meteors. If it were set back in its orbit by ten months' motion, *it would be a part of the meteor swarm.* Similarly, the Perseid meteors have a comet moving in their orbit actually immersed in the stream of meteor particles, and several other of the more conspicuous star showers have comets attending them.

Perhaps the most remarkable case of this character is that of a shower which comes in the latter part of November from the constellation Andromeda, and which from its association with the comet called Biela (after the name of its discoverer) is frequently referred to as the Bielid shower. This comet, an inconspicuous one moving in an unusually small elliptical orbit, had been observed at various times from 1772 down to 1846 without presenting anything remarkable in its appearance; but about the beginning of the latter year, with very little warning, it broke in two, and for three months the pieces were watched by astronomers moving off, side by side, something more than half as far apart as are the earth and moon. It disappeared, made the circuit of its orbit, and six years later came back, with the fragments nearly ten times as far apart as before, and after a short stay near the earth once more disappeared in the distance, never to be seen again, although the fragments should have returned to perihelion at least half a dozen times since then. In one respect the orbit of the comet was remarkable: it passed through the place in which the earth stands on November 27th of each year, so that if the comet were at that particular part of its orbit on any November 27th, a collision between it and the earth would be inevitable. So far as is known, no such collision with the comet has ever occurred, but the Bielid meteors which are strung along its orbit do encounter the earth on that date, in greater or less abundance in different years, and are watched with much interest by the astronomers who look upon them as the final appearance of the *débris* of a worn-out comet.

176. ***Periodic comets.***– The Biela comet is a specimen of the type which astronomers call periodic comets– i. e., those which move in small ellipses and have correspondingly short periodic times, so that they return frequently and regularly to perihelion. The comets which accompany the other meteor swarms– Leonids, Perseids, etc.– also belong to this class as do some 30 or 40 others which have periodic times less than a century. As has been already indicated, these deviations from the normal parabolic orbit call for some special explanation, and the substance of that explanation is contained in the account of the Leonid meteors and their capture by Uranus. Any comet may be thus captured by the attraction of a planet near which it passes. It is only necessary that the perturbing action of the planet should result in a diminution of the comet's velocity, for we have already learned that it is this velocity which determines the character of the orbit, and anything less than the velocity appropriate to a parabola must produce an ellipse– i. e., a closed orbit around which the body will revolve time after time in endless succession. We note in Fig. 115 that when the Leonid swarm encountered Uranus it passed *in front of* the planet and had its velocity diminished and its orbit changed into an ellipse thereby. It might have passed behind Uranus, it would have passed behind had it come a little later, and the effect would then have been just the opposite. Its velocity would have been increased, its orbit changed to a hyperbola, and it would have left the solar system more rapidly than it came into it, thrust out instead of held in by the disturbing planet. Of such cases we can expect no record to remain, but the captured comet is its own witness to what has happened, and bears imprinted upon its orbit the brand of the planet which slowed down its motion. Thus in Fig. 115 the changed orbit of the meteors has its *aphelion* (part remotest from the sun) quite close to the orbit of Uranus, and one of its nodes, ☋, the point in which it cuts through the plane of the ecliptic from north to south side, is also very near to the same orbit. It is these two marks, aphelion and node, which by their position identify Uranus as the planet instrumental in capturing the meteor swarm, and the date of the capture is found by working back with their respective periodic times to an epoch at which planet and comet were simultaneously near this node.

Jupiter, by reason of his great mass, is an especially efficient capturer of comets, and Fig. 116 shows his group of captives, his family of comets as they are sometimes called. The several orbits are marked with the names commonly given to the comets. Frequently this is the name of their discoverer, but often a different system is followed– e. g., the name 1886, IV, means the fourth comet to pass through perihelion in the year 1886. The other great planets– Saturn, Uranus, Neptune– have also their families of captured comets, and according to Schulhof, who does not entirely agree with the common opinion about captured comets, the earth has caught no less than nine of these bodies.

FIG. 116.—Jupiter's family of comets.

177. ***Comet groups.***— But there is another kind of comet family, or comet group as it is called, which deserves some notice, and which is best exemplified by the Great Comet of 1882 and its relatives. No less than four other comets are known to be traveling in substantially the same orbit with this one, the group consisting of comets 1668, I; 1843, I; 1880, I; 1882, II; 1887, I. The orbit itself is not quite a parabola, but a very elongated ellipse, whose major axis and corresponding periodic time can not be very accurately determined from the available data, but it certainly extends far beyond the orbit of Neptune, and requires not less than 500 years for the comet to complete a revolution in it. It was for a time supposed that some one of the recent comets of this group of five might be a return of the comet of 1668 brought back ahead of time by unknown perturbations. There is still a possibility of this, but it is quite out of the question to suppose that the last

four members of the group are anything other than separate and distinct comets moving in practically the same orbit. This common orbit suggests a common origin for the comets, but leaves us to conjecture how they became separated.

The observed orbits of these five comets present some slight discordances among themselves, but if we suppose each comet to move in the average of the observed paths it is a simple matter to fix their several positions at the present time. They have all receded from the sun nearly on line toward the bright star Sirius, and were all of them, at the beginning of the year 1900, standing nearly motionless inside of a space not bigger than the sun and distant from the sun about 150 radii of the earth's orbit. The great rapidity with which they swept through that part of their orbit near the sun (see § 162) is being compensated by the present extreme slowness of their motions, so that the comets of 1668 and 1882, whose passages through the solar system were separated by an interval of more than two centuries, now stand together near the aphelion of their orbits, separated by a distance only 50 per cent greater than the diameter of the moon's orbit, and they will continue substantially in this position for some two or three centuries to come.

The slowness with which these bodies move when far from the sun is strikingly illustrated by an equation of celestial mechanics which for parabolic orbits takes the place of Kepler's Third Law— viz.:

$r^3 / T^2 = 178$,

where T is the time, in years, required for the comet to move from its perihelion to any remote part of the orbit, whose distance from the sun is represented, in radii of the earth's orbit, by r. If the comet of 1668 had moved in a parabola instead of the ellipse supposed above, how many years would have been required to reach its present distance from the sun?

178. ***Relation of comets to the solar system.***— The orbits of these comets illustrate a tendency which is becoming ever more strongly marked. Because comet orbits are nearly parabolas, it used to be assumed that they were exactly parabolic, and this carried with it the conclusion that comets have their origin outside the solar system. It may be so, and this view is in some degree supported by the fact that these nearly parabolic orbits of both comets and meteors are tipped at all possible angles to the plane of the ecliptic instead of lying near it as do the orbits of the planets; and by the further fact that, unlike the planets, the comets show no marked tendency to move around their orbits in the direction in which the sun rotates upon his axis. There is, in fact, the utmost confusion among them in this respect, some going one way and some another. The law of the solar system (gravitation) is impressed upon their movements, but its order is not.

But as observations grow more numerous and more precise, and comet orbits are determined with increasing accuracy, there is a steady gain in the number of elliptic orbits at the expense of the parabolic ones, and if comets are of extraneous origin we must admit that a very considerable percentage of them have their velocities slowed down within the solar system, perhaps not so much by the attraction of the planets as by the resistance offered to their motion by meteor particles and swarms along their paths. A striking instance of what may befall a comet in this way is shown in Fig. 117, where the tail of a comet appears sadly distorted and broken by what is presumed to have been a collision with a meteor swarm. A more famous case of impeded motion is offered by the comet which bears the name of Encke. This has a periodic time less than that of any other known comet, and at intervals of forty months comes back to perihelion, each time moving in a little smaller orbit than before, unquestionably on account of some resistance which it has suffered.

179. ***The development of a comet.***— We saw in § 174 that the sun's action upon a meteor swarm tends to break it up into a long stream, and the same tendency to break up is true of comets whose attenuated substance presents scant resistance to this force. According to the mathematical analysis of Roche, if the comet stood still the sun's tidal force would tend first to draw it out on line with the sun, just as the earth's tidal force pulled the moon out of shape (§ 42), and then it would cause the lighter part of the comet's substance to flow away from both ends of this long diameter. This destructive action of the sun is not limited to comets and meteor streams, for it tends to tear the earth and moon to pieces as well; but the densities and the resulting mutual attractions of their parts are far too great to permit this to be accomplished. As a curiosity of mathematical analysis we may note that a spherical cloud of meteors, or dust particles weighing a gramme each, and placed at the earth's distance from the sun, will be broken up and dissipated by the sun's tidal action if the average distance between the particles exceeds two yards.

Now, the earth is far more dense than such a cloud, whose extreme tenuity, however, suggests what we have already learned of the small density of comets, and prepares us in their case for an outflow of particles at both ends of the diameter directed toward the sun. Something of this kind actually occurs, for the tail of a comet streams out on the side opposite to the sun, and in general points away from the sun, as is shown in Fig. 109, and the envelopes and jets rise up toward the sun; but an inspection of Fig. 106 will show that the tail and the envelope are too unlike to be produced by one and the same set of forces.

It was long ago suggested that the sun possibly exerts upon a comet's substance a repelling force in addition to the attracting force which we call gravity. We think naturally in this connection of the repelling force which a charge of electricity exerts upon a similar charge placed on a neighboring body, and we note that if both sun and comet carried a considerable store of electricity upon their surfaces this would furnish just such a repelling force as seems indicated by the phenomena of comets' tails; for the force of gravity would operate between the substance of sun and comet, and on the whole would be the controlling force, while the electric charges would produce a repulsion, relatively feeble for the big particles and strong for the little ones, since an electric charge lies wholly on the surface, while gravity permeates the whole mass of a body, and the ratio of volume (gravity) to surface (electric charge) increases rapidly with increasing size.

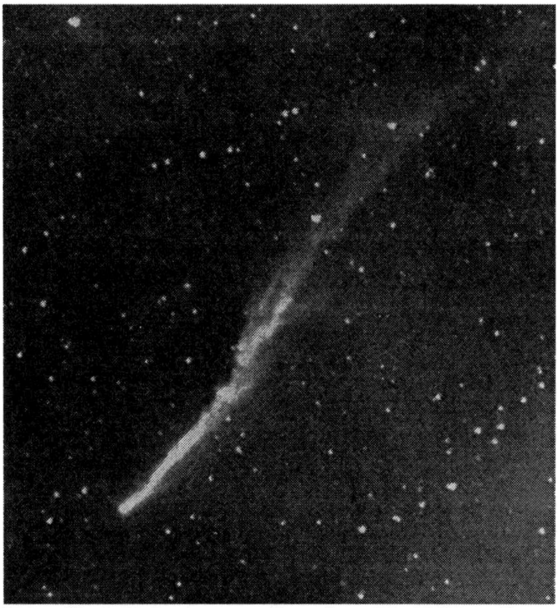

*FIG. 117.— **Brooks's comet, October 21, 1893.**— BARNARD.*

The repelling force would thrust back toward the comet those particles which flowed out toward the sun, while it would urge forward those which flowed away from it, thus producing the difference in appearance between tail and envelopes, the latter being regarded from this standpoint as stunted tails strongly curved backward. In recent years the Russian astronomer Bredichin has made a careful study of the shape and positions of comets' tails and finds that they fit with mathematical precision to the theories of electric repulsion.

180. ***Comet tails.***— According to Bredichin, a comet's tail is formed by something like the following process: In the head of the comet itself a certain part of its matter is broken up into fine bits, single molecules perhaps, which, as they no longer cling together, may be described as in the condition of vapor. By the repellent action of both sun and comet these molecules are cast out from the head of the comet and stream away in the direction opposite to

the sun with different velocities, the heavy ones slowly and the light ones faster, much as particles of smoke stream away from a smokestack, making for the comet a tail which like a trail of smoke is composed of constantly changing particles. The result of this process is shown in Fig. 118, where the positions of the comet in its orbit on successive days are marked by the Roman numerals, and the broken lines represent the paths of molecules m^{I}, m^{II}, m^{III}, etc., expelled from it on their several dates and traveling thereafter in orbits determined by the combined effect of the sun's attraction, the sun's repulsion, and the comet's repulsion. The comet's attraction (gravity) is too small to be taken into account. The line drawn upward from *VI* represents the positions of these molecules on the sixth day, and shows that all of them are arranged in a tail pointing nearly away from the sun. A similar construction for the other dates gives the corresponding positions of the tail, always pointing away from the sun.

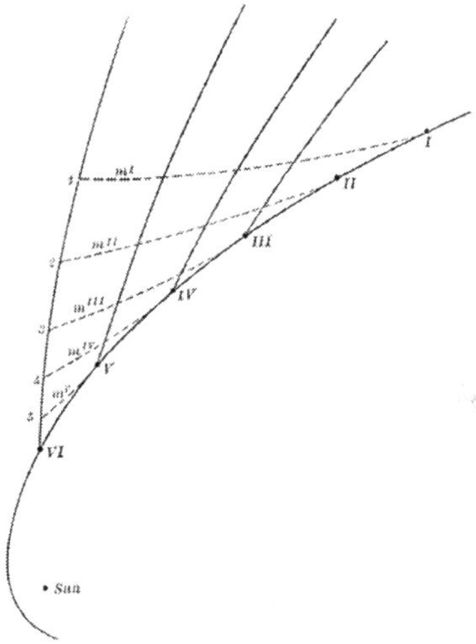

FIG. 118.— Formation of a comet's tail.

Only the lightest kind of molecules— e. g., hydrogen— could drift away from the comet so rapidly as is here shown. The heavier ones, such as carbon and iron, would be repelled as strongly by the electric forces, but they would be more strongly pulled back by the gravitative forces, thus producing a much slower separation between them and the head of the comet. Construct a figure such as the above, in which the molecules shall recede from the comet only one eighth as fast as in Fig. 118, and note what a different position it gives to the comet's tail. Instead of pointing directly away from the sun, it will be bent strongly to one side, as is the large plume-shaped tail of the Donati comet shown in Fig. 101. But observe that this comet has also a nearly straight tail, like the theoretical one of Fig. 118. We have here two distinct types of comet tails, and according to Bredichin there is still another but unusual type, even more strongly bent to one side of the line joining comet and sun, and appearing quite short and stubby. The existence of these three types, and their peculiarities of shape and position, are all satisfactorily accounted for by the supposition that they are made of different materials. The relative molecular weights of hydrogen, some of the hydrocarbons, and iron, are such that tails composed of these molecules would behave just as do the actual tails observed and classified into these three types. The spectroscope shows that these materials– hydrogen, hydrocarbons, and iron– are present in comets, and leaves little room for doubt of the essential soundness of Bredichin's theory.

181. ***Disintegration of comets.*** – We must regard the tail as waste matter cast off from the comet's head, and although the amount of this matter is very small, it must in some measure diminish the comet's mass. This process is, of course, most active at the time of perihelion passage, and if the comet returns to perihelion time after time, as the periodic ones which move in elliptic orbits must do, this waste of material may become a serious matter, leading ultimately to the comet's destruction. It is significant in this connection that the periodic comets are all small and inconspicuous, not one of them showing a tail of any considerable dimensions, and it appears probable that they are far advanced along the road which, in the case of Biela's comet, led to its disintegration. Their fragments are in part strewn through the solar system, making some small fraction of its cloud of cosmic dust, and in part they have been carried away from the sun and scattered throughout the universe along hyperbolic orbits impressed upon them at the time they left the comet.

But it is not through the tail only that the disintegrating process is worked out. While Biela's comet is perhaps the most striking instance in which the head has broken up, it is by no means the only one. The Great Comet of 1882 cast off a considerable number of fragments which moved away as independent though small comets and other more recent comets have been seen to do the same. An even more striking phenomenon was the gradual breaking up of the nucleus of the same comet, 1882, II, into a half dozen nuclei arranged in line like beads upon a string, and pointing along the axis of the tail. See Fig. 119, which shows the series of changes observed in the head of this comet.

182. ***Comets and the spectroscope.*** – The spectrum presented by comets was long a puzzle, and still retains something of that character, although much progress has been made toward an understanding of it. In general it consists of two quite distinct parts– first, a faint background of continuous spectrum due to ordinary sunlight reflected from the comet; and, second, superposed upon this, three bright bands like the carbon band shown

at the middle of Fig. 48, only not so sharply defined. These bands make a discontinuous spectrum quite similar to that given off by compounds of hydrogen and carbon, and of course indicate that a part of the comet's light originates in the body itself, which must therefore be incandescent, or at least must contain some incandescent portions.

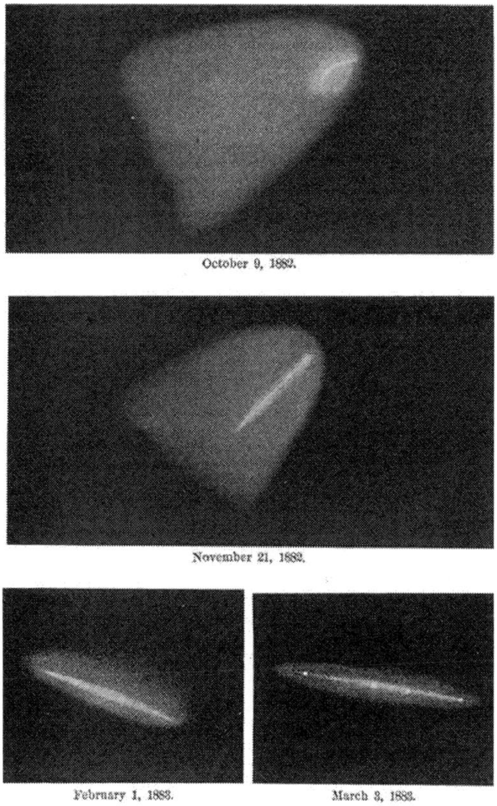

FIG. 119.— The head of the Great Comet of 1882.— WINLOCK.

By heating hydrocarbons in our laboratories until they become incandescent, something like the comet spectrum may be artificially produced, but the best approximation to it is obtained by passing a disruptive electrical discharge through a tube in which fragments of meteors have been placed. A flash of lightning is a disruptive electrical discharge upon a grand scale. Now, meteors and electric phenomena have been independently brought to our notice in connection with comets, and with this suggestion it is easy to frame a general idea of the physical condition of these objects— for example, a cloud of meteors of different sizes so loosely clustered that the average density of the swarm is very low indeed; the several particles in motion relative to each other, as well as to the sun, and disturbed in that motion by the sun's tidal action. Each particle carries its own electric charge, which may be of higher or lower tension than that of its neighbor, and is ready to leap across the intervening gap whenever two particles approach each other. To these conditions add the inductive effect of the sun's electric charge, which tends to produce a particular and artificial distribution of electricity among the comet's particles, and we may expect to find an endless succession of sparks, tiny lightning flashes, springing from one particle to another, most frequent and most vivid when the comet is near the sun, but never strong enough to be separately visible. Their number is, however, great enough to make the comet in part self-luminous with three kinds of light— i. e., the three bright bands of its spectrum, whose wave lengths show in the comet the same elements and compounds of the elements— carbon, hydrogen, and oxygen— which chemical analysis finds in the fallen meteor. It is not to be supposed that these are the only chemical elements in the comet, as they certainly are not the only ones in the meteor. They are the easy ones to detect under ordinary circumstances, but in special cases, like that of the Great Comet of 1882, whose near approach to the sun rendered its whole substance incandescent, the spectrum glows with additional bright lines of sodium, iron, etc.

183. ***Collisions.***— A question sometimes asked, What would be the effect of a collision between the earth and a comet? finds its answer in the results reached in the preceding sections. There would be a star shower, more or less brilliant according to the number and size of the pieces which made up the comet's head. If these were like the remains of the Biela comet, the shower might even be a very tame one; but a collision with a great comet would certainly produce a brilliant meteoric display if its head came in contact with the earth. If the comet were built of small pieces whose individual weights did not exceed a few ounces or pounds, the earth's atmosphere would prove a perfect shield against their attacks, reducing the pieces to harmless dust before they could reach the ground, and leaving the earth uninjured by the encounter, although the comet might suffer sadly from it. But big stones in the comet, meteors too massive to be consumed in their flight through the air, might work a very different effect, and by their bombardment play sad havoc with parts of the earth's surface, although any such result as the wrecking of the earth, or the destruction of all life upon it, does not seem probable. The 40 meteors of § 169 may stand for a collision with a small comet. Consult the Bible (Joshua x, 11) for an example of what might happen with a larger one.

CHAPTER XIII.
THE FIXED STARS

184. ***The constellations.***— In the earlier chapters the student has learned to distinguish between wandering stars (planets) and those fixed luminaries which remain year after year in the same constellation, shining for the most part with unvarying brilliancy, and presenting the most perfect known image of immutability. Homer and Job and prehistoric man saw Orion and the Pleiades much as we see them to-day, although the precession, by changing their relation to the pole of the heavens, has altered their risings and settings, and it may be that their luster has changed in some degree as they grew old with the passing centuries.

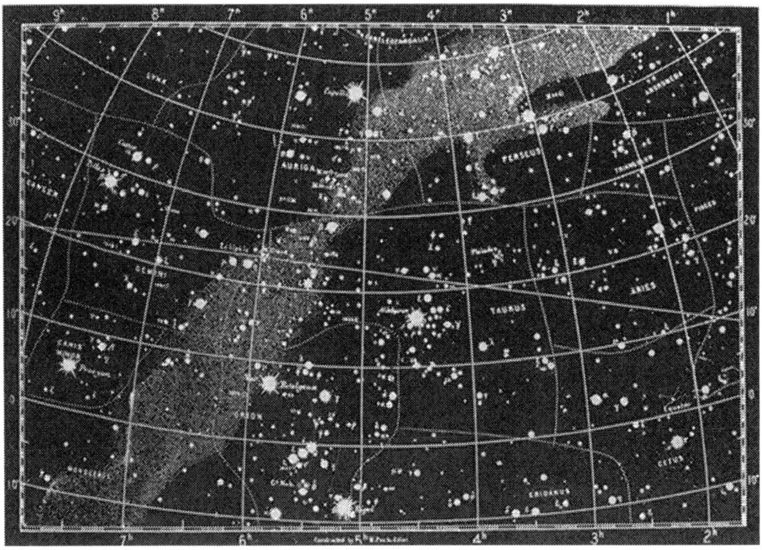

FIG. 120.— Illustrating the division of the sky into constellations.

The division of the sky into constellations dates back to the most primitive times, long before the Christian era, and the crooked and irregular boundaries of these constellations, shown by the dotted lines in Fig. 120, such as no modern astronomer would devise, are an inheritance from antiquity, confounded and made worse in its descent to our day. The boundaries assigned to constellations near the south pole are much more smooth and regular, since this part of the sky, invisible to the peoples from whom we inherit, was not studied and mapped until more modern times. The old traditions associated with each constellation a figure, often drawn from classical mythology, which was supposed to be suggested by the grouping of the stars: thus Ursa Major is a great bear, stalking across the sky, with the handle of the Dipper for his tail; Leo is a lion; Cassiopeia, a lady in a chair; Andromeda, a maiden chained to a rock, etc.; but for the most part the resemblances are far-fetched and quite too fanciful to be followed by the ordinary eye.

185. ***The number of stars.*** – "As numerous as the stars of heaven" is a familiar figure of speech for expressing the idea of countless number, but as applied to the visible stars of the sky the words convey quite a wrong impression, for, under ordinary circumstances, in a clear sky every star to be seen may be counted in the course of a few hours, since they do not exceed 3,000 or 4,000, the exact number depending upon atmospheric conditions and the keenness of the individual eye. Test your own vision by counting the stars of the Pleiades. Six are easily seen, and you may possibly find as many as ten or twelve; but however many are seen, there will be a vague impression of more just beyond the limit of visibility, and doubtless this impression is partly responsible for the popular exaggeration of the number of the stars. In fact, much more than half of what we call starlight comes from stars which are separately too small to be seen, but whose number is so great as to more than make up for their individual faintness.

The Milky Way is just such a cloud of faint stars, and the student who can obtain access to a small telescope, or even an opera glass, should not fail to turn it toward the Milky Way and see for himself how that vague stream of light breaks up into shining points, each an independent star. These faint stars, which are found in every part of the sky as well as in the Milky Way, are usually called *telescopic*, in recognition of the fact that they can be seen only in the telescope, while the other brighter ones are known as *lucid stars.*

186. ***Magnitudes.*** – The telescopic stars show among themselves an even greater range of brightness than do the lucid ones, and the system of magnitudes (§ 9) has accordingly been extended to include them, the faintest star visible in the greatest telescope of the present time being of the sixteenth or seventeenth magnitude, while, as we have already learned, stars on the dividing line between the telescopic and the lucid ones are of the sixth magnitude. To compare the amount of light received from the stars with that from the planets, and particularly from the sun and moon, it has been found necessary to prolong the scale of magnitudes backward into the negative numbers, and we speak of the sun as having a stellar magnitude represented by the number -26.5. The full moon's stellar magnitude is -12, and the planets range from -3 (Venus) to +8 (Neptune). Even a very few of the stars are so bright that negative magnitudes must be used to represent their true relation to the fainter ones. Sirius, for example, the brightest of the fixed stars, is of the -1 magnitude, and such stars as Arcturus and Vega are of the 0 magnitude.

The relation of these magnitudes to each other has been so chosen that a star of any one magnitude is very approximately 2.5 times as bright as one of the next fainter magnitude, and this ratio furnishes a convenient method of comparing the amount of light received from different stars. Thus the brightness of Venus is 2.5 × 2.5 times that of Sirius. The full moon is $(2.5)^9$ times as bright as Venus, etc.; only it should be observed that the number 2.5 is not exactly the value of the *light ratio* between two consecutive magnitudes. Strictly this ratio is the $\sqrt[5]{100} = 2.5119+$, so that to be entirely accurate we must say that a difference of five magnitudes gives a hundredfold difference of brightness. In mathematical symbols, if B represents the ratio of brightness (quantity of light) of two stars whose magnitudes are m and n, then

$$B = (100)^{(m-n)/5}$$

How much brighter is an ordinary first-magnitude star, such as Aldebaran or Spica, than a star just visible to the naked eye? How many of the faintest stars visible in a great telescope would be required to make one star just visible to the unaided eye? How many full moons must be put in the sky in order to give an illumination as bright as daylight? How large a part of the visible hemisphere would they occupy?

187. ***Classification by magnitudes.*** — The brightness of all the lucid stars has been carefully measured with an instrument (photometer) designed for that special purpose, and the following table shows, according to the Harvard Photometry, the number of stars in the whole sky, from pole to pole, which are brighter than the several magnitudes named in the table:

The number	of stars	brighter	than	magnitude	1.0	is	11
"	"	"	"	"	2.0	"	39
"	"	"	"	"	3.0	"	142
"	"	"	"	"	4.0	"	463
"	"	"	"	"	5.0	"	1,483
"	"	"	"	"	6.0	"	4,326

It must not be inferred from this table that there are in the whole sky only 4,326 stars visible to the naked eye. The actual number is probably 50 or 60 per cent greater than this, and the normal human eye sees stars as faint as the magnitude 6.4 or 6.5, the discordance between this number and the previous statement, that the sixth magnitude is the limit of the naked-eye vision, having been introduced in the attempt to make precise and accurate a classification into magnitudes which was at first only rough and approximate. This same striving after accuracy leads to the introduction of fractional numbers to represent gradations of brightness intermediate between whole magnitudes. Thus of the 2,843 stars included between the fifth and sixth magnitudes a certain proportion are said to be of the 5.1 magnitude, 5.2 magnitude, and so on to the 5.9 magnitude, even hundredths of a magnitude being sometimes employed.

We have found the number of stars included between the fifth and sixth magnitudes by subtracting from the last number of the preceding table the number immediately preceding it, and similarly we may find the number included between each other pair of consecutive magnitudes, as follows:

Magnitude	0	1	2	3	4	5	6
Number of stars		1	8	103	321	1,020	2,843
4×3^m		2	6	108	324	972	2,916

In the last line each number after the first is found by multiplying the preceding one by 3, and the approximate agreement of each such number with that printed above it shows that on the whole, as far as the table goes, the fainter stars are approximately three times as numerous as those a magnitude brighter.

The magnitudes of the telescopic stars have not yet been measured completely, and their exact number is unknown; but if we apply our principle of a threefold increase for each successive magnitude, we shall find for the fainter stars— those of the tenth and twelfth magnitudes— prodigious numbers which run up into the millions, and even these are probably too small, since down to the ninth or tenth magnitude it is certain that the number of the telescopic stars increases from magnitude to magnitude in more than a threefold ratio. This is balanced in some degree by the less rapid increase which is known to exist in magnitudes still fainter; and applying our formula without regard to these variations in the rate of increase, we obtain as a rude approximation to the total number of stars down to the fifteenth magnitude, 86,000,000. The Herschels, father and son, actually counted the number of stars visible in nearly 8,000 sample regions of the sky, and, inferring the character of the whole sky from these samples, we find it to contain 58,500,000 stars; but the magnitude of the faintest star visible in their telescope, and included in their count, is rather uncertain.

How many first-magnitude stars would be needed to give as much light as do the 2,843 stars of magnitude 5.0 to 6.0? How many tenth-magnitude stars are required to give the same amount of light?

To the modern man it seems natural to ascribe the different brilliancies of the stars to their different distances from us; but such was not the case 2,000 years ago, when each fixed star was commonly thought to be fastened to a "crystal sphere," which carried them with it, all at the same distance from us, as it turned about the earth. In breaking away from this crroneous idea and learning to think of the sky itself as only an atmospheric illusion through which we look to stars at very different distances beyond, it was easy to fall into the opposite error and to think of the stars as being much alike one with another, and, like pebbles on the beach, scattered throughout space with some rough degree of uniformity, so that in every direction there should be found in equal measure stars near at hand and stars far off, each shining with a luster proportioned to its remoteness.

188. ***Distances of the stars.***– Now, in order to separate the true from the false in this last mode of thinking about the stars, we need some knowledge of their real distances from the earth, and in seeking it we encounter what is perhaps the most delicate and difficult problem in the whole range of observational astronomy. As shown in Fig. 121, the principles involved in determining these distances are not fundamentally different from those employed in determining the moon's distance from the earth. Thus, the ellipse at the left of the figure represents the earth's orbit and the position of the earth at different times of the year. The direction of the star *A* at these several times is shown by lines drawn through *A* and prolonged to the background apparently furnished by the sky. A similar construction is made for the star *B*, and it is readily seen that owing to the changing position of the observer as he moves around the earth's orbit, both *A* and *B* will appear to move upon the background in orbits shaped like that of the earth as seen from the star, but having their size dependent upon the star's distance, the apparent orbit of *A* being larger than that of *B*, because *A* is nearer the earth. By measuring the angular distance between *A* and *B* at opposite seasons of the year (e. g., the angles *A—Jan.—B*, and *A—July—B*) the astronomer determines from the change in this angle how much larger is the one path than the other, and thus concludes how much nearer is *A* than *B*. Strictly, the difference between the January and July angles is equal to the difference between the angles subtended at *A* and *B* by the diameter of the earth's orbit, and if *B* were so far away that the angle *Jan.—B—July* were nothing at all we should get immediately from the observations the angle *Jan.—A—July*, which would suffice to determine the stars' distance. Supposing the diameter of the earth's orbit and the angle at *A* to be known, can you make a graphical construction that will determine the distance of *A* from the earth?

The angle subtended at *A* by the radius of the earth's orbit– i. e., 1/2 (*Jan.—A—July*)– is called the star's parallax, and this is commonly used by astronomers as a measure of the star's distance instead of expressing it in linear units such as miles or radii of the earth's orbit. The distance of a star is equal to the radius of the earth's orbit divided by the parallax, in seconds of arc, and multiplied by the number 206265.

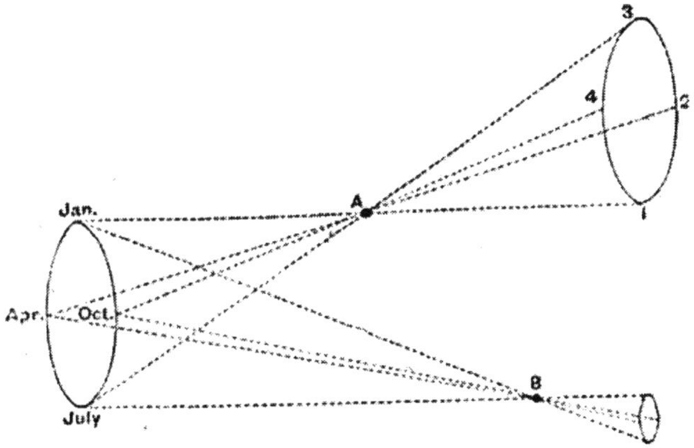

FIG. 121.— Determining a star's parallax.

A weak point of this method of measuring stellar distances is that it always gives what is called a relative parallax— i. e., the difference between the parallaxes of *A* and *B*; and while it is customary to select for *B* a star or stars supposed to be much farther off than *A*, it may happen, and sometimes does happen, that these comparison stars as they are called are as near or nearer than *A*, and give a negative parallax— i. e., the difference between the angles at *A* and *B* proves to be negative, as it must whenever the star *B* is nearer than *A*.

The first really successful determinations of stellar parallax were made by Struve and Bessel a little prior to 1840, and since that time the distances of perhaps 100 stars have been measured with some degree of reliability, although the parallaxes themselves are so small— never as great as 1"— that it is extremely difficult to avoid falling into error, since even for the nearest star the problem of its distance is equivalent to finding the distance of an object more than 5 miles away by looking at it first with one eye and then with the other. Too short a base line.

189. ***The sun and his neighbors.***— The distances of the sun's nearer neighbors among the stars are shown in Fig. 122, where the two circles having the sun at their center represent distances from it equal respectively to 1,000,000 and 2,000,000 times the distance between earth and sun. In the figure the direction of each star from the sun corresponds to its right ascension, as shown by the Roman numerals about the outer circle; the true direction of the star from the sun can not, of course, be shown upon the flat surface of the paper, but it may be found by elevating or depressing the star from the surface of the paper through an angle, as seen from the sun, equal to its declination, as shown in the fifth column of the following table,

The Sun's Nearest Neighbors

No.	*STAR.*	*Magnitude.*	*R. A.*	*Dec.*	*Parallax.*	*Distance.*
1	α Centauri	0.7	14.5h.	-60°	0.75"	0.27
2	Ll. 21,185	6.8	11.0	+37	0.45	0.46
3	61 Cygni	5.0	21.0	+38	0.40	0.51
4	η Herculis	3.6	16.7	+39	0.40	0.51
5	Sirius	-1.4	6.7	-17	0.37	0.56
6	Σ 2,398	8.2	18.7	+59	0.35	0.58
7	Procyon	0.5	7.6	+5	0.34	0.60
8	γ Draconis	4.8	17.5	+55	0.30	0.68
9	Gr. 34	7.9	0.2	+43	0.29	0.71
10	Lac. 9,352	7.5	23.0	-36	0.28	0.74
11	σ Draconis	4.8	19.5	+69	0.25	0.82
12	A. O. 17,415-6	9.0	17.6	+68	0.25	0.82
13	η Cassiopeiæ	3.4	0.7	+57	0.25	0.82
14	Altair	1.0	19.8	+9	0.21	0.97

No.	STAR.	Magnitude.	R. A.	Dec.	Parallax.	Distance.
15	ϵ Indi	5.2	21.9	-57	0.20	1.03
16	Gr. 1,618	6.7	10.1	+50	0.20	1.03
17	10 Ursæ Majoris	4.2	8.9	+42	0.20	1.03
18	Castor	1.5	7.5	+32	0.20	1.03
19	Ll. 21,258	8.5	11.0	+44	0.20	1.03
20	o^2 Eridani	4.5	4.2	-8	0.19	1.08
21	A. O. 11,677	9.0	11.2	+66	0.19	1.08
22	Ll. 18,115	8.0	9.1	+53	0.18	1.14
23	B. D. 36°, 3,883	7.1	20.0	+36	0.18	1.14
24	Gr. 1,618	6.5	10.1	+50	0.17	1.21
25	β Cassiopeiæ	2.3	0.1	+59	0.16	1.28
26	70 Ophiuchi	4.4	18.0	+2	0.16	1.28
27	Σ 1,516	6.5	11.2	+74	0.15	1.38
28	Gr. 1,830	6.6	11.8	+39	0.15	1.38
29	μ Cassiopeiæ	5.4	1.0	+54	0.14	1.47

No.	*Star.*	*Magnitude.*	*R. A.*	*Dec.*	*Parallax.*	*Distance.*
30	ϑ Eridani	4.4	3.5	-10	0.14	1.47
31	ι Ursæ Majoris	3.2	8.9	+48	0.13	1.58
32	β Hydri	2.9	0.3	-78	0.1	1.58
33	Fomalhaut	1.0	22.9	-30	0.13	1.58
34	Br. 3,077	6.0	23.1	+57	0.13	1.58
35	ϑ Cygni	2.5	20.8	+33	0.12	1.71
36	β Comæ	4.5	13.1	+28	0.11	1.87
37	ψ^5 Aurigæ	8.8	6.6	+44	0.11	1.87
38	π Herculis	3.3	17.2	+37	0.11	1.87
39	Aldebaran	1.1	4.5	+16	0.10	2.06
40	Capella	0.1	5.1	+46	0.10	2.06
41	B. D. 35°, 4,003	9.2	20.1	+35	0.10	2.06
42	Gr. 1,646	6.3	10.3	+49	0.10	2.06
43	γ Cygni	2.3	20.3	+40	0.10	2.06
44	Regulus	1.2	10.0	+12	0.10	2.06

No.	*STAR.*	*Magnitude.*	*R. A.*	*Dec.*	*Parallax.*	*Distance.*
45	Vega	0.2	18.6	+39	0.10	2.06

in which the numbers in the first column are those placed adjacent to the stars in the diagram to identify them.

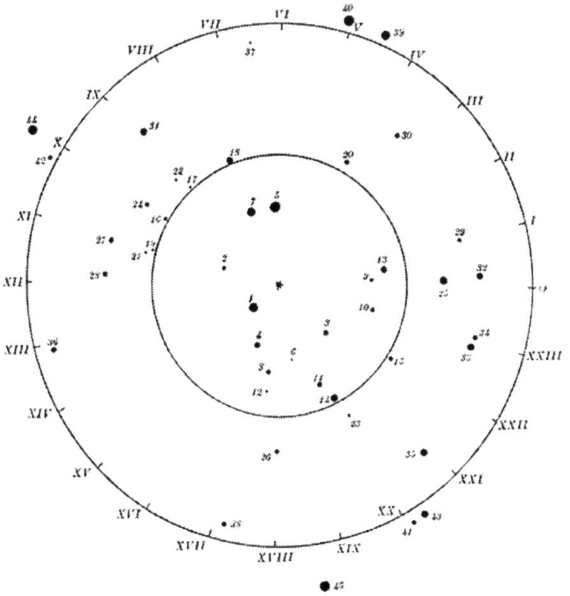

FIG. 122.—Stellar neighbors of the sun.

190. ***Light years.***— The radius of the inner circle in Fig. 122, 1,000,000 times the earth's distance from the sun, is a convenient unit in which to express the stellar distances, and in the preceding table the distances of the stars from the sun are expressed in terms of this unit. To express them in miles the numbers in the table must be multiplied by 93,000,000,000,000. The nearest star, α Centauri, is 25,000,000,000,000 miles away. But there is another unit in more common use— i. e., the distance traveled over by light in the period of one year. We have already found (§ 141) that it requires light 8m. 18s. to come from the sun to the earth, and it is a simple matter to find from this datum that in a year light moves over a space equal to 63,368 radii of the earth's orbit. This distance is called a *light year*, and the distance of the same star, α Centauri, expressed in terms of this unit, is 4.26 years— i. e., it takes light that long to come from the star to the earth.

In Fig. 122 the stellar magnitudes of the stars are indicated by the size of the dots– the bigger the dot the brighter the star – and a mere inspection of the figure will serve to show that within a radius of 30 light years from the sun bright stars and faint ones are mixed up together, and that, so far as distance is concerned, the sun is only a member of this swarm of stars, whose distances apart, each from its nearest neighbor, are of the same order of magnitude as those which separate the sun from the three or four stars nearest it.

Fig. 122 is not to be supposed complete. Doubtless other stars will be found whose distance from the sun is less than 2,000,000 radii of the earth's orbit, but it is not probable that they will ever suffice to more than double or perhaps treble the number here shown. The vast majority of the stars lie far beyond the limits of the figure.

191. ***Proper motions.*** – It is evident that these stars are too far apart for their mutual attractions to have much influence one upon another, and that we have here a case in which, according to § 34, each star is free to keep unchanged its state of rest or motion with unvarying velocity along a straight line. Their very name, *fixed stars*, implies that they are at rest, and so astronomers long believed. Hipparchus (125 B. C.) and Ptolemy (130 A. D.) observed and recorded many allineations among the stars, in order to give to future generations a means of settling this very question of a possible motion of the stars and a resulting change in their relative positions upon the sky. For example, they found at the beginning of the Christian era that the four stars, Capella, ϑ Persei, α and β Arietis, stood in a straight line– i. e., upon a great circle of the sky. Verify this by direct reference to the sky, and see how nearly these stars have kept the same position for nearly twenty centuries. Three of them may be identified from the star maps, and the fourth, ϑ Persei, is a third-magnitude star between Capella and the other two.

Other allineations given by Ptolemy are: Spica, Arcturus and β Bootis; Spica, δ Corvi and γ Corvi; α Libræ, Arcturus and ζ Ursæ Majoris. Arcturus does not now fit very well to these alignments, and nearly two centuries ago it, together with Aldebaran and Sirius, was on other grounds suspected to have changed its place in the sky since the days of Ptolemy. This discovery, long since fully confirmed, gave a great impetus to observing with all possible accuracy the right ascensions and declinations of the stars, with a view to finding other cases of what was called *proper motion*– i. e., a motion peculiar to the individual star as contrasted with the change of right ascension and declination produced for all stars by the precession.

Since the middle of the eighteenth century there have been made many thousands of observations of this kind, whose results have gone into star charts and star catalogues, and which are now being supplemented by a photographic survey of the sky that is intended to record permanently upon photographic plates the position and magnitude of every star in the heavens down to the fourteenth magnitude, with a view to ultimately determining all their proper motions.

The complete achievement of this result is, of course, a thing of the remote future, but sufficient progress in determining these motions has been made during the past century and a half to show that nearly every lucid star possesses some proper motion, although in most cases it is very small, there being less than 100 known stars in which it amounts to so much as 1" per annum– i. e., a rate of motion across the sky which would require nearly the whole Christian era to alter a star's direction from us by so much as the moon's angular diameter. The most rapid known proper motion is that of a telescopic star midway between the equator and the south pole, which changes its position at the rate of nearly 9" per annum, and the next greatest is that of another telescopic star, in the northern sky, No. 28 of Fig. 122. It is not until we reach the tenth place in a list of large proper motions that we find a bright lucid star, No. 1 of Fig. 122. It is a significant fact that for the most part the stars with large proper motions are precisely the ones shown in Fig. 122, which is designed to show stars near the earth. This connection between nearness and rapidity of proper motions is indeed what we should expect to find, since a given amount of real motion of the star along its orbit will produce a larger angular displacement, proper motion, the nearer the star is to the earth, and this fact has guided astronomers in selecting the stars to be observed for parallax, the proper motion being determined first and the parallax afterward.

192. ***The paths of the stars.*** – We have already seen reason for thinking that the orbit along which a star moves is practically a straight line, and from a study of proper motions, particularly their directions across the sky, it appears that these orbits point in all possible ways– north, south, east, and west– so that some of them are doubtless directed nearly toward or from the sun; others are square to the line joining sun and star; while the vast majority occupy some position intermediate between these two. Now, our relation to these real motions of the stars is well illustrated in Fig. 112, where the observer finds in some of the shooting stars a tremendous proper motion across the sky, but sees nothing of their rapid approach to him, while others appear to stand motionless, although, in fact, they are moving quite as rapidly as are their fellows. The fixed star resembles the shooting star in this respect, that its proper motion is only that part of its real motion which lies at right angles to the line of sight, and this needs to be supplemented by that other part of the motion which lies parallel to the line of sight, in order to give us any knowledge of the star's real orbit.

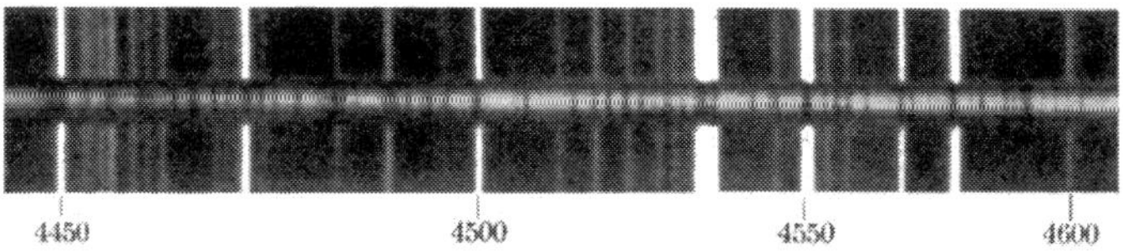

FIG. 123.— Motion of Polaris in the line of sight as determined by the spectroscope. FROST.

193. ***Motion in the line of sight.***— It is only within the last 25 years that anything whatever has been accomplished in determining these stellar motions of approach or recession, but within that time much progress has been made by applying the Doppler principle (§ 89) to the study of stellar spectra, and at the present time nearly every great telescope in the world is engaged upon work of this kind. The shifting of the lines of the spectrum toward the violet or toward the red end of the spectrum indicates with certainty the approach or recession of the star, but this shifting, which must be determined by comparing the star's spectrum with that of some artificial light showing corresponding lines, is so small in amount that its accurate measurement is a matter of extreme difficulty, as may be seen from Fig. 123. This cut shows along its central line a part of the spectrum of Polaris, between wave lengths 4,450 and 4,600 tenth meters, while above and below are the corresponding parts of the spectrum of an electric spark whose light passed through the same spectroscope and was photographed upon the same plate with that of Polaris. This comparison spectrum is, as it should be, a discontinuous or bright-line one, while the spectrum of the star is a continuous one, broken only by dark gaps or lines, many of which have no corresponding lines in the comparison spectrum. But a certain number of lines in the two spectra do correspond, save that the dark line is always pushed a very little toward the direction of shorter wave lengths, showing that this star is approaching the earth. This spectrum was photographed for the express purpose of determining the star's motion in the line of sight, and with it there should be compared Figs. 124 and 125, which show in the upper part of each a photograph obtained without comparison spectra by allowing the star's light to pass through some prisms placed just in front of the telescope. The lower section of each figure shows an enlargement of the original photograph, bringing out its details in a way not visible to the unaided eye. In the enlarged spectrum of β Aurigæ a rate of motion equal to that of the earth in its orbit would be represented by a shifting of 0.03 of a millimeter in the position of the broad, hazy lines.

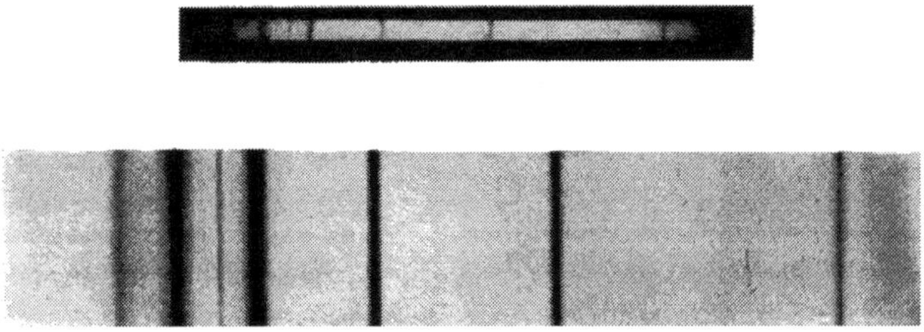

FIG. 124.— Spectrum of β Aurigæ.— PICKERING.

Despite the difficulty of dealing with such small quantities as the above, very satisfactory results are now obtained, and from them it is known that the velocities of stars in the line of sight are of the same order of magnitude as the velocities of the planets in their orbits, ranging all the way from 0 to 60 miles per second— more than 200,000 miles per hour— which latter velocity, according to Campbell, is the rate at which μ Cassiopeiæ is approaching the sun.

The student should not fail to note one important difference between proper motions and the motions determined spectroscopically: the latter are given directly in miles per second, or per hour, while the former are expressed in angular measure, seconds of arc, and there can be no direct comparison between the two until by means of the known distances of the stars their proper motions are converted from angular into linear measure. We are brought thus to the very heart of the matter; parallax, proper motion, and motion in the line of sight are intimately related quantities, all of which are essential to a knowledge of the real motions of the stars.

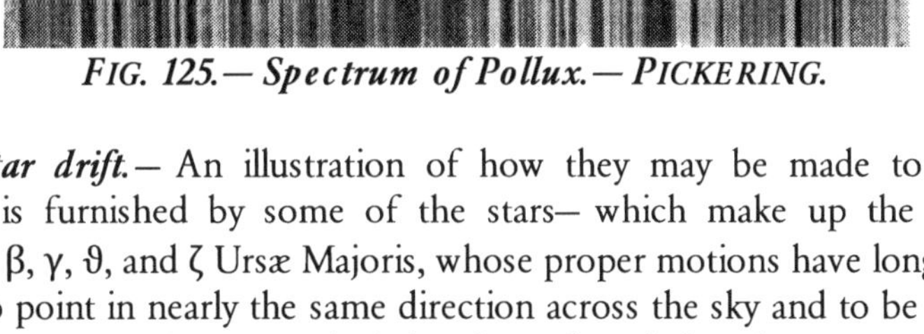

FIG. 125.— Spectrum of Pollux.— PICKERING.

194. ***Star drift.***— An illustration of how they may be made to work together is furnished by some of the stars— which make up the Great Dipper— β, γ, ϑ, and ζ Ursæ Majoris, whose proper motions have long been known to point in nearly the same direction across the sky and to be nearly equal in amount. More recently it has been found that these stars are all moving toward the sun with approximately the same velocity— 18 miles per second. One other star of the Dipper, δ Ursæ Majoris, shares in the common proper motion, but its velocity in the line of sight has not yet been determined

with the spectroscope. These similar motions make it probable that the stars are really traveling together through spacc along parallel lines; and on the supposition that such is the case it is quite possible to write out a set of equations which shall involve their known proper motions and motions in the line of sight, together with their unknown distances and the unknown direction and velocity of their real motion along their orbits. Solving these equations for the values of the unknown quantities, it is found that the five stars probably lie in a plane which is turned nearly edgewise toward us, and that in this plane they are moving about twice as fast as the earth moves around the sun, and are at a distance from us represented by a parallax of less than 0.02"— i. e., six times as great as the outermost circle in Fig. 122. A most extraordinary system of stars which, although separated from each other by distances as great as the whole breadth of Fig. 122, yet move along in parallel paths which it is difficult to regard as the result of chance, and for which it is equally difficult to frame an explanation.

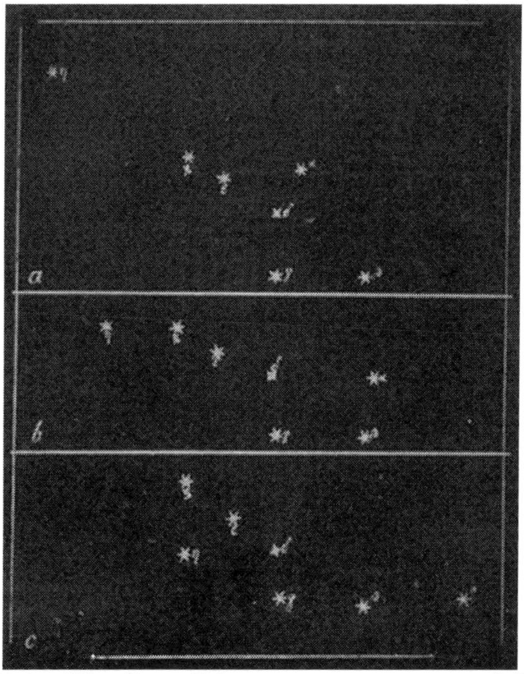

FIG. 126.— The Great Dipper, past, present, and future.

The stars α and η of the Great Dipper do not share in this motion, and must ultimately part company with the other five, to the complete destruction of the Dipper's shape. Fig. 126 illustrates this change of shape, the upper part of the figure (*a*) showing these seven stars as they were grouped at a remote epoch in the past, while the lower section (*c*) shows their position for an equally remote epoch in the future. There is no resemblance to a dipper in either of these configurations, but it should be observed that in each of them the stars α and η keep their relative position unaltered, and the other five stars also keep together, the entire change of appearance being due to the changing positions of these two groups with respect to each other.

This phenomenon of groups of stars moving together is called *star drift*, and quite a number of cases of it are found in different parts of the sky. The Pleiades are perhaps the most conspicuous one, for here some sixty or more stars are found traveling together along similar paths. Repeated careful measurements of the relative positions of stars in this cluster show that one of the lucid stars and four or five of the telescopic ones do not share in this motion, and therefore are not to be considered as members of the group, but rather as isolated stars which, for a time, chance to be nearly on line with the Pleiades, and probably farther off, since their proper motions are smaller.

To rightly appreciate the extreme slowness with which proper motions alter the constellations, the student should bear in mind that the changes shown in passing from one section of Fig. 126 to the next represent the effect of the present proper motions of the stars accumulated for a period of 200,000 years. Will the stars continue to move in straight paths for so long a time?

195. ***The sun's way.***— Another and even more interesting application of proper motions and motions in the line of sight is the determination from them of the sun's orbit among the stars. The principle involved is simple enough. If the sun moves with respect to the stars and carries the earth and the other planets year after year into new regions of space, our changing point of view must displace in some measure every star in the sky save those which happen to be exactly on the line of the sun's motion, and even these will show its effect by their apparent motion of approach or recession along the line of sight. So far as their own orbital motions are concerned, there is no reason to suppose that more stars move north than south, or that more go east than west; and when we find in their proper motions a distinct tendency to radiate from a point somewhere near the bright star Vega and to converge toward a point on the opposite side of the sky, we infer that this does not come from any general drift of the stars in that direction, but that it marks the course of the sun among them. That it is moving along a straight line pointing toward Vega, and that at least a part of the velocities which the spectroscope shows in the line of sight, comes from the motion of the sun and earth. Working along these lines, Kapteyn finds that the sun is moving through space with a velocity of 11 miles per second, which is decidedly below the average rate of stellar motion— 19 miles per second.

196. ***Distance of Sirian and solar stars.***— By combining this rate of motion of the sun with the average proper motions of the stars of different

magnitudes, it is possible to obtain some idea of the average distance from us of a first-magnitude star or a sixth-magnitude star, which, while it gives no information about the actual distance of any particular star, does show that on the whole the fainter stars are more remote. But here a broad distinction must be drawn. By far the larger part of the stars belong to one of two well-marked classes, called respectively Sirian and solar stars, which are readily distinguished from each other by the kind of spectrum they furnish. Thus β Aurigæ belongs to the Sirian class, as does every other star which has a spectrum like that of Fig. 124, while Pollux is a solar star presenting in Fig. 125 a spectrum like that of the sun, as do the other stars of this class.

Two thirds of the sun's near neighbors, shown in Fig. 122, have spectra of the solar type, and in general stars of this class are nearer to us than are the stars with spectra unlike that of the sun. The average distance of a solar star of the first magnitude is very approximately represented by the outer circle in Fig. 122, 2,000,000 times the distance of the sun from the earth; while the corresponding distance for a Sirian star of the first magnitude is represented by the number 4,600,000.

A third-magnitude star is on the average twice as far away as one of the first magnitude, a fifth-magnitude star four times as far off, etc., each additional two magnitudes doubling the average distance of the stars, at least down to the eighth magnitude and possibly farther, although beyond this limit we have no certain knowledge. Put in another way, the naked eye sees many Sirian stars which *may* have "gone out" and ceased to shine centuries ago, for the light by which we now see them left those stars before the discovery of America by Columbus. For the student of mathematical tastes we note that the results of Kapteyn's investigation of the mean distances (D) of the stars of magnitude (m) may be put into two equations:

For Solar Stars, $$D = 23 \times 2^{m/2}$$

For Sirian Stars, $$D = 52 \times 2^{m/2}$$

where the coefficients 23 and 52 are expressed in light years. How long a time is required for light to come from an average solar star of the sixth magnitude?

197. ***Consequences of stellar distance.***— The amount of light which comes to us from any luminous body varies inversely as the square of its distance, and since many of the stars are changing their distance from us quite rapidly, it must be that with the lapse of time they will grow brighter or fainter by reason of this altered distance. But the distances themselves are so great that the most rapid known motion in the line of sight would require more than 1,000 years (probably several thousand) to produce any perceptible change in brilliancy.

The law in accordance with which this change of brilliancy takes place is that the distance must be increased or diminished tenfold in order to produce a change of five magnitudes in the brightness of the object, and we may apply this law to determine the sun's rank among the stars. If it were removed to the distance of an average first-, or second-, or third-magnitude star, how would its light compare with that of the stars? The average distance of a third-magnitude star of the solar type is, as we have seen above, 4,000,000 times the sun's distance from the earth, and since $4,000,000 = 10^{6.6}$, we find that at this distance the sun's stellar magnitude would be altered by 6.6×5 magnitudes, and would therefore be $-26.5 + 33.0 = 6.5$– i. e., the sun if removed to the average distance of the third-magnitude stars of its type would be reduced to the very limit of naked-eye visibility. It must therefore be relatively small and feeble as compared with the brightness of the average star. It is only its close proximity to us that makes the sun look brighter than the stars.

The fixed stars may have planets circling around them, but an application of the same principles will show how hopeless is the prospect of ever seeing them in a telescope. If the sun's nearest neighbor, α Centauri, were attended by a planet like Jupiter, this planet would furnish to us no more light than does a star of the twenty-second magnitude– i. e., it would be absolutely invisible, and would remain invisible in the most powerful telescope yet built, even though its bulk were increased to equal that of the sun. Let the student make the computation leading to this result, assuming the stellar magnitude of Jupiter to be -1.7.

198. ***Double stars.***– In the constellation Taurus, not far from Aldebaran, is the fourth-magnitude star θ Tauri, which can readily be seen to consist of two stars close together. The star α Capricorni is plainly double, and a sharp eye can detect that one of the faint stars which with Vega make a small equilateral triangle, is also a double star. Look for them in the sky.

In the strict language of astronomy the term double star would not be applied to the first two of these objects, since it is usually restricted to those stars whose angular distance from each other is so small that in the telescope they appear much as do the stars named above to the naked eye– i. e., their angular separation is measured by a few seconds or fractions of a single second, instead of the six minutes which separate the component stars of θ Tauri or α Capricorni. There are found in the sky many thousands of these close double stars, of which some are only optically double– i. e., two stars nearly on line with the earth but at very different distances from it– while more of them are really what they seem, stars near each other, and in many cases near enough to influence each other's motion. These are called *binary* systems, and in cases of this kind the principles of celestial mechanics set forth in Chapter IV hold true, and we may expect to find each component of a double star moving in a conic section of some kind, having its focus at the common center of gravity of the two stars. We are thus presented with problems of orbital motion quite similar to those which occur in the solar system, and careful telescopic observations are required year after year to fix the relative positions of the two stars– i. e., their angular separation, which it is customary to call their *distance*, and their direction one from the other, which is called *position angle*.

199. ***Orbits of double stars.***– The sun's nearest neighbor, α Centauri, is such a double star, whose position angle and distance have been measured by successive generations of astronomers for more than a century, and Fig. 127 shows the result of plotting their observations. Each black dot that lies on or near the circumference of the long ellipse stands for an observed direction and distance of the fainter of the two stars from the brighter one, which is represented by the small circle at the intersection of the lines inside the ellipse. It appears from the figure that during this time the one star has gone completely around the other, as a planet goes around the sun, and the true orbit must therefore be an ellipse having one of its foci at the center of gravity of the two stars. The other star moves in an ellipse of precisely similar shape, but probably smaller size, since the dimensions of the two orbits are inversely proportional to the masses of the two bodies, but it is customary to neglect this motion of the larger star and to give to the smaller one an orbit whose diameter is equal to the sum of the diameters of the two real orbits. This practice, which has been followed in Fig. 127, gives correctly the relative positions of the two stars, and makes one orbit do the work of two.

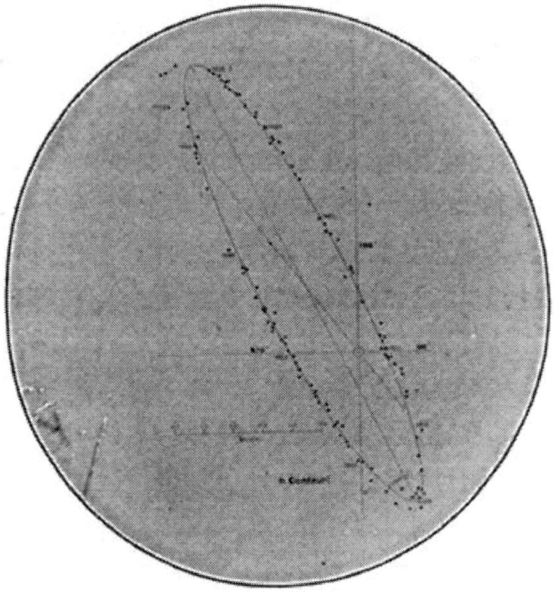

FIG. 127.— The orbit of α Centauri.— SEE.

In Fig. 127 the bright star does not fall anywhere near the focus of the ellipse marked out by the smaller one, and from this we infer that the figure does not show the true shape of the orbit, which is certainly distorted, foreshortened, by the fact that we look obliquely down upon its plane. It is possible, however, by mathematical analysis, to find just how much and in what direction that plane should be turned in order to bring the focus of the ellipse up to the position of the principal star, and thus give the true shape and size of the orbit. See Fig. 128 for a case in which the true orbit is turned exactly edgewise toward the earth, and the small star, which really moves in an ellipse like that shown in the figure, appears to oscillate to and fro along a straight line drawn through the principal star, as shown at the left of the figure.

In the case of α Centauri the true orbit proves to have a major axis 47 times, and a minor axis 40 times, as great as the distance of the earth from the sun. The orbit, in fact, is intermediate in size between the orbits of Uranus and Neptune, and the periodic time of the star in this orbit is 81 years, a little less than the period of Uranus.

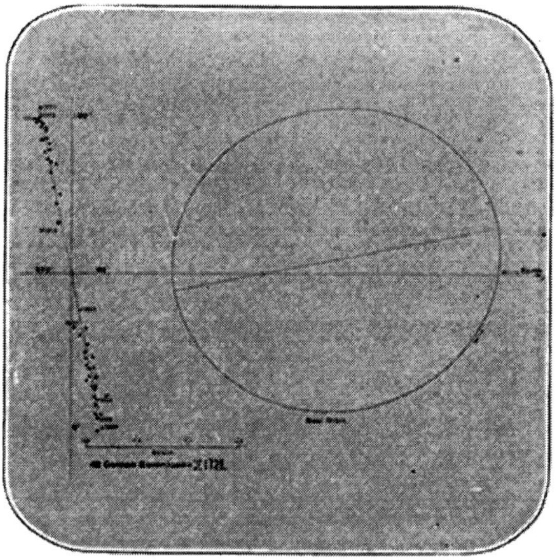

FIG. 128.— Apparent orbit and real orbit of the double star 42 Comæ Berenicis.— SEE.

200. ***Masses of double stars.***— If we apply to this orbit Kepler's Third Law in the form given it at page 179, we shall find—

$$a^3 / T^2 = (23.5)^3 / (81)^2 = k (M + m),$$

where M and m represent the masses of the two stars. We have already seen that k, the gravitation constant, is equal to 1 when the masses are measured in terms of the sun's mass taken as unity, and when T and a are expressed in years and radii of the earth's orbit respectively, and with this value of k we may readily find from the above equation, $M + m = 2.5$– i. e., the combined mass of the two components of α Centauri is equal to rather more than twice the mass of the sun. It is not every double star to which this process of weighing can be applied. The major axis of the orbit, a, is found from the observations in angular measure, 35" in this case, and it is only when the parallax of the star is known that this can be converted into the required linear units, radii of the earth's orbit, by dividing the angular major axis by the parallax; 47 = 35" ÷ 0.75".

Our list of distances (§ 189) contains four double stars whose periodic times and major axes have been fairly well determined, and we find in the accompanying table the information which they give about the masses of double stars and the size of the orbits in which they move:

STAR.	*Major axis.*	*Minor axis.*	*Periodic time.*	*Mass.*
α Centauri	47	40	81 y.	2
70 Ophiuchi	56	48	88	3
Procyon	34	31	40	3
Sirius	43	34	52	4

The orbit of Uranus, diameter = 38, and Neptune, diameter = 60, are of much the same size as these double-star orbits; but the planetary orbits are nearly circular, while in every case the double stars show a substantial difference between the long and short diameters of their orbits. This is a characteristic feature of most double-star orbits, and seems to stand in some relation to their periodic times, for, on the average, the longer the time required by a star to make its orbital revolution the more eccentric is its orbit likely to prove.

Another element of the orbits of double stars, which stands in even closer relation to the periodic time, is the major axis; the smaller the long diameter of the orbit the more rapid is the motion and the shorter the periodic time, so that astronomers in search of interesting double-star orbits devote themselves by preference to those stars whose distance apart is so small that they can barely be distinguished one from the other in the telescope.

Although the half-dozen stars contained in the table all have orbits of much the same size and with much the same periodic time as those in which Uranus and Neptune move, this is by no means true of all the double stars, many of which have periods running up into the hundreds if not thousands of years, while a few complete their orbital revolutions in periods comparable with, or even shorter than, that of Jupiter.

201. ***Dark stars.*** — Procyon, the next to the last star of the preceding table, calls for some special mention, as the determination of its mass and orbit stands upon a rather different basis from that of the other stars. More than half a century ago it was discovered that its proper motion was not straight and uniform after the fashion of ordinary stars, but presented a series of loops like those marked out by a bright point on the rim of a swiftly running bicycle wheel. The hub may move straight forward with uniform velocity, but the point near the tire goes up and down, and, while sharing in the forward motion of the hub, runs sometimes ahead of it, sometimes behind, and such seemed to be the motion of Procyon and of Sirius as well. Bessel, who discovered it, did not hesitate to apply the laws of motion, and to affirm that this visible change of the star's motion pointed to the presence of an unseen companion, which produced upon the motions of Sirius and Procyon just such effects as the visible companions produce in the motions of double stars. A new kind of star, dark instead of bright, was added to the astronomer's domain, and its discoverer boldly suggested the possible existence of many more. "That countless stars are visible is clearly no argument against the existence of as many more invisible ones." "There is no reason to think radiance a necessary property of celestial bodies." But most astronomers were incredulous, and it was not until 1862 that, in the testing of a new and powerful telescope just built, a dark star was brought to light and the companion of Sirius actually seen. The visual discovery of the dark companion of Procyon is of still more recent date (November, 1896), when it was detected with the great telescope of the Lick Observatory. This discovery is so recent that the orbit is still very uncertain, being based almost wholly upon the variations in the proper motion of the star, and while the periodic time must be very nearly correct, the mass of the stars and dimensions of the orbit may require considerable correction.

The companion of Sirius is about ten magnitudes and that of Procyon about twelve magnitudes fainter than the star itself. How much more light does the bright star give than its faint companion? Despite the tremendous difference of brightness represented by the answer to this question, the mass of Sirius is only about twice as great as that of its companion, and for Procyon the ratio does not exceed five or six.

The visual discovery of the companions to Sirius and Procyon removes them from the list of dark stars, but others still remain unseen, although their existence is indicated by variable proper motions or by variable orbital motion, as in the case of ζ Cancri, where one of the components of a triple star moves around the other two in a series of loops whose presence indicates a disturbing body which has never yet been seen.

202. ***Multiple stars.*** – Combinations of three, four, or more stars close to each other, like ζ Cancri, are called multiple stars, and while they are far from being as common as are double stars, there is a considerable number of them in the sky, 100 or more as against the more than 10,000 double stars that are known. That their relative motions are subject to the law of gravitation admits of no serious doubt, but mathematical analysis breaks down in face of the difficulties here presented, and no astronomer has ever been able to determine what will be the general character of the motions in such a system.

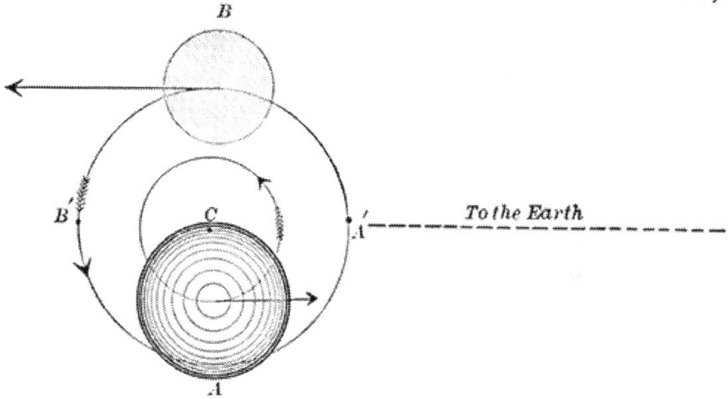

FIG. 129.– Illustrating the motion of a spectroscopic binary.

203. ***Spectroscopic binaries.*** – In the year 1890 Professor Pickering, of the Harvard Observatory, announced the discovery of a new class of double stars, invisible as such in even the most powerful telescope, and producing no perturbations such as have been considered above, but showing in their spectrum that two or more bodies must be present in the source of light which to the eye is indistinguishable from a single star. In Fig. 129 we suppose *A* and *B* to be the two components of a double star, each moving in its own orbit about their common center of gravity, *C*, whose distance from the earth is several million times greater than the distance between the stars themselves. Under such circumstances no telescope could distinguish between the two stars, which would appear fused into one; but the smaller the orbit the more rapid would be their motion in it, and if this orbit were turned edgewise toward the earth, as is supposed in the figure, whenever the stars were in the relative position there shown, *A* would be rapidly approaching the earth by reason of its orbital motion, while *B* would move away from it, so that in accordance with the Doppler principle the lines composing their respective spectra would be shifted in opposite directions, thus producing a doubling of the lines, each single line breaking up into two, like the double-sodium line *D*, only not spaced so far apart. When the stars have moved a quarter way round their orbit to the points *A'*, *B'*, their

velocities are turned at right angles to the line of sight and the spectrum returns to the normal type with single lines, only to break up again when after another quarter revolution their velocities are again parallel with the line of sight. The interval of time between consecutive doublings of the lines in the spectrum thus furnishes half the time of a revolution in the orbit. The distance between the components of a double line shows by means of the Doppler principle how fast the stars are traveling, and this in connection with the periodic times fixes the size of the orbit, provided we assume that it is turned exactly edgewise to the earth. This assumption may not be quite true, but even though the orbit should deviate considerably from this position, it will still present the phenomenon of the double lines whose displacement will now show something less than the true velocities of the stars in their orbits, since the spectroscope measures only that component of the whole velocity which is directed toward the earth, and it is important to note that the real orbits and masses of these *spectroscopic binaries*, as they are called, will usually be somewhat larger than those indicated by the spectroscope, since it is only in exceptional cases that the orbit will be turned exactly edgewise to us.

The bright star Capella is an excellent illustration of these spectroscopic binaries. At intervals of a little less than a month the lines of its spectrum are alternately single and double, their maximum separation corresponding to a velocity in the line of sight amounting to 37 miles per second. Each component of a doubled line appears to be shifted an equal amount from the position occupied by the line when it is single, thus indicating equal velocities and equal masses for the two component stars whose periodic time in their orbit is 104 days. From this periodic time, together with the velocity of the star's motion, let the student show that the diameter of the orbit— i. e., the distance of the stars from each other— is approximately 53,000,000 miles, and that their combined mass is a little less than that of α Centauri, provided that their orbit plane is turned exactly edgewise toward the earth.

There are at the present time (1901) 34 spectroscopic binaries known, including among them such stars as Polaris, Capella, Algol, Spica, β Aurigæ, ζ Ursæ Majoris, etc., and their number is rapidly increasing, about one star out of every seven whose motion in the line of sight is determined proving to be a binary or, as in the case of Polaris, possibly triple. On account of smaller distance apart their periodic times are much shorter than those of the ordinary double stars, and range from a few days up to several months— more than two years in the case of η Pegasi, which has the longest known period of any star of this class.

Spectroscopic binaries agree with ordinary double stars in having masses rather greater than that of the sun, but there is as yet no assured case of a mass ten times as great as that of the sun.

204. ***Variable stars.***— Attention has already been drawn (§ 23) to the fact that some stars shine with a changing brightness— e. g., Algol, the most

famous of these *variable stars*, at its maximum of brightness furnishes three times as much light as when at its minimum, and other variable stars show an even greater range. The star o Ceti has been named Mira (Latin, *the wonderful*), from its extraordinary range of brightness, more than six-hundred-fold. For the greater part of the time this star is invisible to the naked eye, but during some three months in every year it brightens up sufficiently to be seen, rising quite rapidly to its maximum brilliancy, which is sometimes that of a second-magnitude star, but more frequently only third or even fourth magnitude, and, after shining for a few weeks with nearly maximum brilliancy, falling off to become invisible for a time and then return to its maximum brightness after an interval of eleven months from the preceding maximum. In 1901 it should reach its greatest brilliancy about midsummer, and a month earlier than this for each succeeding year. Find it by means of the star map, and by comparing its brightness from night to night with neighboring stars of about the same magnitude see how it changes with respect to them.

The interval of time from maximum to maximum of brightness— 331.6 days for Mira— is called the star's period, and within its period a star regularly variable runs through all its changes of brilliancy, much as the weather runs through its cycle of changes in the period of a year. But, as there are wet years and dry ones, hot years and cold, so also with variable stars, many of them show differences more or less pronounced between different periods, and one such difference has already been noted in the case of Mira; its maximum brilliancy is different in different years. So, too, the length of the period fluctuates in many cases, as does every other circumstance connected with it, and predictions of what such a variable star will do are notoriously unreliable.

205. ***The Algol variables.***— On the other hand, some variable stars present an almost perfect regularity, repeating their changes time after time with a precision like that of clockwork. Algol is one type of these regular variables, having a period of 68.8154 hours, during six sevenths of which time it shines with unchanging luster as a star of the 2.3 magnitude, but during the remaining 9 hours of each period it runs down to the 3.5 magnitude, and comes back again, as is shown by a curve in Fig. 130. The horizontal scale here represents hours, reckoned from the time of the star's minimum brightness, and the vertical scale shows stellar magnitudes. Such a diagram is called the star's light curve, and we may read from it that at any time between 5h. and 32h. after the time of minimum the star's magnitude is 2.32; at 2h. after a minimum the magnitude is 2.88, etc. What is the magnitude an hour and a half before the time of minimum? What is the magnitude 43 days after a minimum?

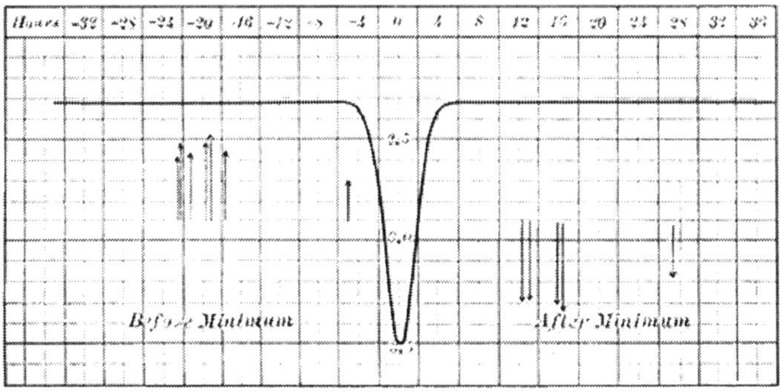

FIG. 130.— The light curve of Algol.

The arrows shown in Fig. 130 are a feature not usually found with light curves, but in this case each one represents a spectroscopic determination of the motion of Algol in the line of sight. These observations extended over a period of more than two years, but they are plotted in the figure with reference to the number of hours each one preceded or followed a minimum of the star's light, and each arrow shows not only the direction of the star's motion along the line of sight, the arrows pointing down denoting approach of the star toward the earth, but also its velocity, each square of the ruling corresponding to 10 kilometers (6.2 miles per second). The differences of velocity shown by adjacent arrows come mainly from errors of observation and furnish some idea of how consistent among themselves such observations are, but there can be no doubt that before minimum the star is moving away from the earth, and after minimum is approaching it. It is evident from these observations that in Algol we have to do with a spectroscopic binary, one of whose components is a dark star which, once in each revolution, partially eclipses the bright star and produces thus the variations in its light. By combining the spectroscopic observations with the variations in the star's light, Vogel finds that the bright star, Algol, itself has a diameter somewhat greater than that of the sun, but is of low density, so that its mass is less than half that of the sun, while the dark star is a very little smaller than the sun and has about a quarter of its mass. The distance between the two stars, dark and bright, is 3,200,000 miles. Fig. 129, which is drawn to scale, shows the relative positions and sizes of these stars as well as the orbits in which they move.

The mere fact already noted that close binary systems exist in considerable numbers is sufficient to make it probable that a certain proportion of these stars would have their orbit planes turned so nearly edgewise toward the earth as to produce eclipses, and corresponding to this probability there are already known no less than 15 stars of the Algol type of eclipse variables, and only a beginning has been made in the search for them.

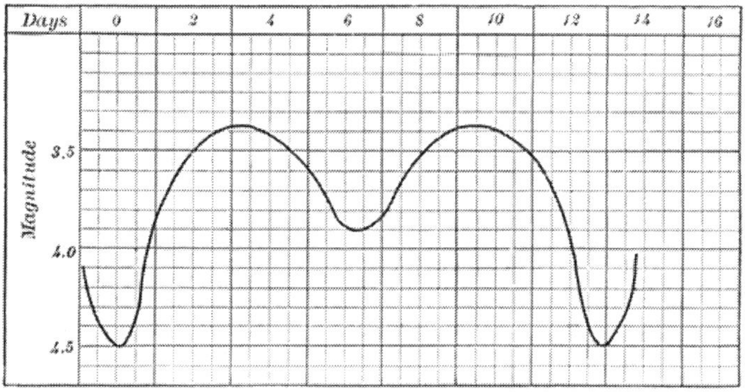

FIG. 131.— The light curve of β Lyræ.

206. ***Variables of the β Lyræ type.***— In addition to these there is a certain further number of binary variables in which both components are bright and where the variation of brightness follows a very different course. Capella would be such a variable if its orbit plane were directed exactly toward the earth, and the fact that its light is not variable shows conclusively that such is not the position of the orbit. Fig. 131 represents the light curve of one of the best-known variable systems of this second type, that of β Lyræ, whose period is 12 days 21.8 hours, and the student should read from the curve the magnitude of the star for different times during this interval. According to Myers, this light curve and the spectroscopic observations of the star point to the existence of a binary star of very remarkable character, such as is shown, together with its orbit and a scale of miles, in Fig. 132. Note the tide which each of these stars raises in the other, thus changing their shapes from spheres into ellipsoids. The astonishing dimensions of these stars are in part compensated by their very low density, which is less than that of air, so that their masses are respectively only 10 times and 21 times that of the sun! But these dimensions and masses perhaps require confirmation, since they depend upon spectroscopic observations of doubtful interpretation. In Fig. 132 what relative positions must the stars occupy in their orbit in order that their combined light should give β Lyræ its maximum brightness? What position will furnish a minimum brightness?

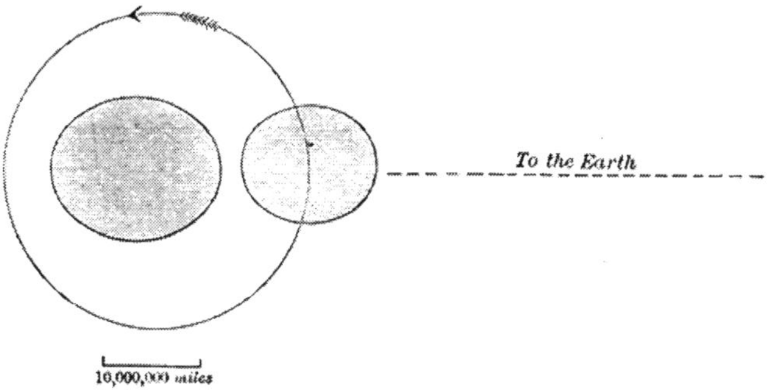

FIG. 132.— The system of β Lyræ.— MYERS.

207. ***Variables of long and short periods.***— It must not be supposed that all variable stars are binaries which eclipse each other. By far the larger part of them, like Mira, are not to be accounted for in this way, and a distinction which is pretty well marked in the length of their periods is significant in this connection. There is a considerable number of variable stars with periods shorter than a month, and there are many having periods longer than 6 months, but there are very few having periods longer than 18 months, or intermediate between 1 month and 6 months, so that it is quite customary to divide variable stars into two classes— those of long period, 6 months or more, and those of short period less than 6 months, and that this distinction corresponds to some real difference in the stars themselves is further marked by the fact that the long-period variables are prevailingly red in color, while the short-period stars are almost without exception white or very pale yellow. In fact, the longer the period the redder the star, although it is not to be inferred that all red stars are variable; a considerable percentage of them shine with constant light. The eclipse explanation of variability holds good only for short-period variables, and possibly not for all of them, while for the long-period variables there is no explanation which commands the general assent of astronomers, although unverified hypotheses are plenty.

The number of stars known to be variable is about 400, while a considerable number of others are "suspected," and it would not be surprising if a large fraction of all the stars should be found to fluctuate a little in brightness. The sun's spots may suffice to make it a variable star with a period of 11 years.

The discovery of new variables is of frequent occurrence, and may be expected to become more frequent when the sky is systematically explored for them by the ingenious device suggested by Pickering and illustrated in Fig. 133. A given region of the sky— e. g., the Northern Crown— is photographed repeatedly upon the same plate, which is shifted a little at each new exposure, so that the stars shall fall at new places upon it. The finally developed plate shows a row of images corresponding to each star, and if the star's light is constant the images in any given row will all be of the same size, as are most of those in Fig. 133; but a variable star such as is shown by the arrowhead reveals its presence by the broken aspect of its row of dots, a minimum brilliancy being shown by smaller and a maximum by larger ones. In this particular case, at two exposures the star was too faint to print its image upon the plate.

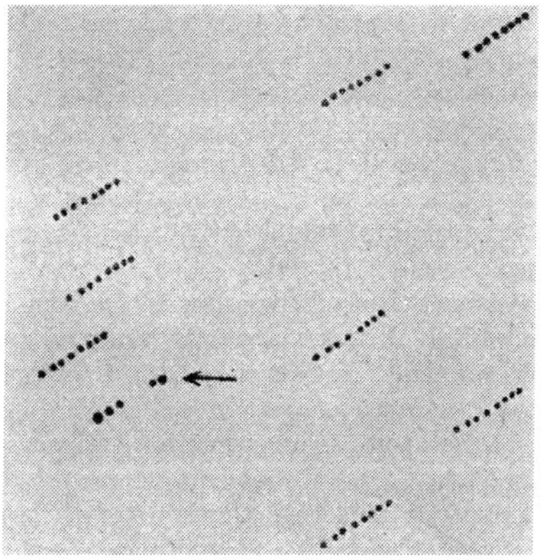

FIG. 133.— Discovery of a variable star by means of photography.— PICKERING.

208. ***New stars.***— Next to the variable stars of very long or very irregular period stand the so-called *new* or *temporary stars*, which appear for the most part suddenly, and after a brief time either vanish altogether or sink to comparative insignificance. These were formerly thought to be very remarkable and unusual occurrences— "the birth of a new world"— and it is noteworthy that no new star is recorded to have been seen from 1670 to 1848 A. D., for since that time there have been no less than five of them visible to the naked eye and others telescopic. In so far as these new stars are not ordinary variables (Mira, first seen in 1596, was long counted as a new star), they are commonly supposed due to chance encounters between stars or other cosmic bodies moving with considerable velocities along orbits which approach very close to each other. The actual collision of two dark bodies moving with high velocities is clearly sufficient to produce a luminous star— e. g., meteors— and even the close approach of two cooled-off stars, might result in tidal actions which would rend open their crusts and pour out the glowing matter from within so as to produce temporarily a very great accession of brightness.

The most famous of all new stars is that which, according to Tycho Brahe's report, appeared in the year 1572, and was so bright when at its best as to be seen with the naked eye in broad daylight. It continued visible, though with fading light, for about 16 months, and finally disappeared to the naked eye, although there is some reason to suppose that it can be identified with a ruddy star of the eleventh magnitude in the constellation Cassiopeia, whose light still shows traces of variability.

No modern temporary star approaches that of Tycho in splendor, but in some respects the recent ones surpass it in interest, since it has been possible to apply the spectroscope to the analysis of their light and to find thereby a much more complex set of conditions in the star than would have been suspected from its light changes alone.

One of the most extraordinary of new stars, and the most brilliant one since that of Tycho, appeared suddenly in the constellation Perseus in February, 1901, and for a short time equaled Capella in brightness. But its light rapidly waned, with periodic fluctuations of brightness like those of a variable star, and at the present time (September, 1902) it is lost to the naked eye, although in the telescope it still shines like a star of the ninth or tenth magnitude.

By the aid of powerful photographic apparatus, during the period of its waning brilliancy a ring of faint nebulous matter was detected surrounding the star and drifting around and away from it much as if a series of nebulæ had been thrown off by the star at the time of its sudden outburst of light. But the extraordinary velocity of this nebular motion, nearly a billion miles per hour, makes such an explanation almost incredible, and astronomers are more inclined to believe that the ring was merely a reflection of the star's own light from a cloud of meteoric matter, into which a rapidly moving dark star plunged and, after the fashion of terrestrial meteors, was raised to brilliant incandescence by the collision. If we assume this to be the true explanation of these extraordinary phenomena, it is possible to show from the known velocity with which light travels through space and from the rate at which the nebula spread, that the distance of Nova Persei, as the new star is called, corresponds to a parallax of about one one-hundredth of a second, a result that is, in substance, confirmed by direct telescopic measurements of its parallax.

Another modern temporary star is Nova Aurigæ, which appeared suddenly in December, 1891, waned, and in the following April vanished, only to reappear three months later for another season of renewed brightness. The spectra of both these modern Novæ contain both dark and bright lines displaced toward opposite ends of the spectrum, and suggesting the Doppler effect that would be produced by two or more glowing bodies having rapid and opposite motions in the line of sight. But the most recent investigations cast discredit on this explanation and leave the spectra of temporary stars still a subject of debate among astronomers, with respect both to the motion they indicate and the intrinsic nature of the stars themselves. The varying aspect of the spectra suggested at one time the sun's chromosphere, at another time the conditions that are present in nebulæ, etc.

CHAPTER XIII.
THE FIXED STARS

CHAPTER XIV.
STARS AND NEBULÆ

209. ***Stellar colors.***— We have already seen that one star differs from another in respect of color as well as brightness, and the diligent student of the sky will not fail to observe for himself how the luster of Sirius and Rigel is more nearly a pure white than is that of any other stars in the heavens, while at the other end of the scale α Orionis and Aldebaran are strongly ruddy, and Antares presents an even deeper tone of red. Between these extremes the light of every star shows a mixture of the rainbow hues, in which a very pale yellow is the predominant color, shading off, as we have seen, to white at one end of the scale and red at the other. There are no green stars, or blue stars, or violet stars, save in one exceptional class of cases— viz., where the two components of a double star are of very different brightness, it is quite the usual thing for them to have different colors, and then, almost without exception, the color of the fainter star lies nearer to the violet end of the spectrum than does the color of the bright one, and sometimes shows a distinctly blue or green hue. A fine type of such double star is β Cygni, in which the components are respectively yellow and blue, and the yellow star furnishes eight times as much light as the blue one.

The exception which double stars thus make to the general rule of stellar colors, yellow and red, but no color of shorter wave length, has never been satisfactorily explained, but the rule itself presents no difficulties. Each star is an incandescent body, giving off radiant energy of every wave length within the limits of the visible spectrum, and, indeed, far beyond these limits. If this radiant energy could come unhindered to our eyes every star would appear white, but they are all surrounded by atmospheres— analogous to the chromosphere and reversing layer of the sun— which absorb a portion of their radiant energy and, like the earth's atmosphere, take a heavier toll from the violet than from the red end of the spectrum. The greater the absorption in the star's atmosphere, therefore, the feebler and the ruddier will be its light, and corresponding to this the red stars are as a class fainter than the white ones.

210. ***Chemistry of the stars.***— The spectroscope is pre-eminently the instrument to deal with this absorption of light in the stellar atmospheres, just as it deals with that absorption in the sun's atmosphere to which are due the dark lines of the solar spectrum, although the faintness of starlight, compared with that of the sun, presents a serious obstacle to its use. Despite this difficulty most of the lucid stars and many of the telescopic ones have been studied with the spectroscope and found to be similar to the sun and the earth as respects the material of which they are made. Such familiar chemical elements as hydrogen and iron, carbon, sodium, and calcium are scattered broadcast throughout the visible universe, and while it would be unwarranted

by the present state of knowledge to say that the stars contain nothing not found in the earth and the sun, it is evident that in a broad way their substance is like rather than unlike that composing the solar system, and is subject to the same physical and chemical laws which obtain here. Galileo and Newton extended to the heavens the terrestrial sciences of mathematics and mechanics, but it remained to the nineteenth century to show that the physics and chemistry of the sky are like the physics and chemistry of the earth.

211. ***Stellar spectra.***– When the spectra of great numbers of stars are compared one with another, it is found that they bear some relation to the colors of the stars, as, indeed, we should expect, since spectrum and color are both produced by the stellar atmospheres, and it is found useful to classify these spectra into three types, as follows:

Type I. Sirian stars.– Speaking generally, the stars which are white or very faintly tinged with yellow, furnish spectra like that of Sirius, from which they take their name, or that of β Aurigæ (Fig. 124), which is a continuous spectrum, especially rich in energy of short wave length– i. e., violet and ultra-violet light, and is crossed by a relatively small number of heavy dark lines corresponding to the spectrum of hydrogen. Sometimes, however, these lines are much fainter than is here shown, and we find associated with them still other faint ones pointing to the presence of other metallic substances in the star's atmosphere. These metallic lines are not always present, and sometimes even the hydrogen lines themselves are lacking, but the spectrum is always rich in violet and ultra-violet light.

Since with increasing temperature a body emits a continually increasing proportion of energy of short wave length (§ 118), the richness of these spectra in such energy points to a very high temperature in these stars, probably surpassing in some considerable measure that of the sun. Stars with this type of spectrum are more numerous than all others combined, but next to them in point of numbers stands–

Type II. Solar stars.– To this type of spectrum belong the yellow stars, which show spectra like that of the sun, or of Pollux (Fig. 125). These are not so rich in violet light as are those of Type I, but in complexity of spectrum and in the number of their absorption lines they far surpass the Sirian stars. They are supposed to be at a lower temperature than the Sirian stars, and a much larger number of chemical elements seems present and active in the reversing layer of their atmospheres. The strong resemblance which these spectra bear to that of the sun, together with the fact that most of the sun's stellar neighbors have spectra of this type, justify us in ranking both them and it as members of one class, called *solar stars.*

Type III. Red stars.– A small number of stars show spectra comparable with that of α Herculis (Fig. 134), in which the blue and the violet part of the spectrum is almost obliterated, and the remaining yellow and red parts show not only dark lines, but also numerous broad dark bands, sharp at one edge, and gradually fading out at the other. It is this *selective absorption*, extinguishing the blue and leaving the red end of the spectrum, which produces the ruddy color of these stars, while the bands in their spectra "are characteristic of chemical combinations, and their presence ... proves that at certain elevations in the atmospheres of these stars the temperature has sunk so low that chemical combinations can be formed and maintained" (Scheiner-Frost). One of the chemical compounds here indicated is a hydrocarbon similar to that found in comets. In the white and yellow stars the temperatures are so high that the same chemical elements, although present, can not unite one with another to form compound substances.

FIG. 134.— The spectrum of α Herculis.— ESPIN.

Most of the variable stars are red and have spectra of the third type; but this does not hold true for the eclipse variables like Algol, all of which are white stars with spectra of the first type. The ordinary variable star is therefore one with a dense atmosphere of relatively low temperature and complex structure, which produces the prevailing red color of these stars by absorbing the major part of their radiant energy of short wave length while allowing the longer, red waves to escape. Although their exact nature is not understood, there can be little doubt that the fluctuation in the light of these stars is due to processes taking place within the star itself, but whether above or below its photosphere is still uncertain.

212. ***Classes of stars.***— There is no hard-and-fast dividing line between these types of stellar spectra, but the change from one to another is by insensible gradations, like the transition from youth to manhood and from manhood to old age, and along the line of transition are to be found numberless peculiarities and varieties of spectra not enumerated above— e. g., a few stars show not only dark absorption lines in their spectra but bright lines as well, which, like those in Fig. 48, point to the presence of incandescent vapors, even in the outer parts of their atmospheres. Among the lucid stars about 75 per cent have spectra of the first type, 23 per cent are of the second type, 1 per cent of the third type, and the remaining 1 per cent are peculiar or of doubtful classification. Among the telescopic stars it is probable that much the same distribution holds, but in the present state of knowledge it is not prudent to speak with entire confidence upon this point.

That the great number of stars whose spectra have been studied should admit of a classification so simple as the above, is an impressive fact which, when supplemented by the further fact of a gradual transition from one type of spectrum to the next, leaves little room for doubt that in the stars we have an innumerable throng of individuals belonging to the same species but in different stages of development, and that the sun is only one of these individuals, of something less than medium size and in a stage of development which is not at all peculiar, since it is shared by nearly a fourth of all the stars.

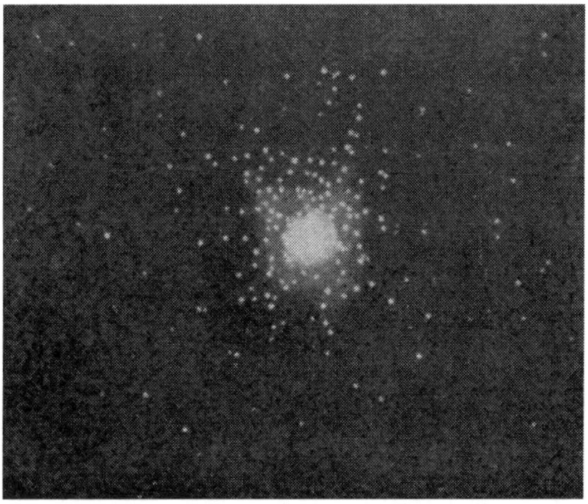

FIG 135.— Star cluster in Hercules.

213. ***Star clusters.***— In previous chapters we have noted the Pleiades and Præsepe as star clusters visible to the naked eye, and to them we may add the Hyades, near Aldebaran, and the little constellation Coma Berenices. But more impressive than any of these, although visible only in a telescope, is the splendid cluster in Hercules, whose appearance in a telescope of moderate size is shown in Fig. 135, while Fig. 136 is a photograph of the same cluster taken with a very large reflecting telescope. This is only a type of many telescopic clusters which are scattered over the sky, and which are made up of stars packed so closely together as to become indistinguishable, one from another, at the center of the cluster. Within an area which could be covered by a third of the full moon's face are crowded in this cluster more than five thousand stars which are unquestionably close neighbors, but whose apparent nearness to each other is doubtless due to their great distance from us. It is quite probable that even at the center of this cluster, where more than a thousand stars are included within a radius of 160", the actual distances separating adjoining stars are much greater than that separating earth and sun, but far less than that separating the sun from its nearest stellar neighbor.

An interesting discovery of recent date, made by Professor Bailey in photographing star clusters, is that some few of them, which are especially rich in stars, contain an extraordinary number of variable stars, mostly very faint and of short period. Two clusters, one in the northern and one in the southern hemisphere, contain each more than a hundred variables, and an even more extraordinary case is presented by a cluster, called Messier 5, not far from the star α Serpentis, which contains no less than sixty-three variables, all about of the fourteenth magnitude, all having light periods which differ but little from half a day, all having light curves of about the same shape, and all having a range of brightness from maximum to minimum of about one magnitude. An extraordinary set of coincidences which "points unmistakably to a common origin and cause of variability."

FIG. 136.—Star cluster in Hercules.—KEELER.

FIG. 137.— The Andromeda nebula as seen in a very small telescope.

FIG. 138.— The Andromeda nebula and Holmes's comet. Photographed by BARNARD.

FIG. 139.—A drawing of the Andromeda nebula.

FIG. 140.—A photograph of the Andromeda nebula.—ROBERTS.

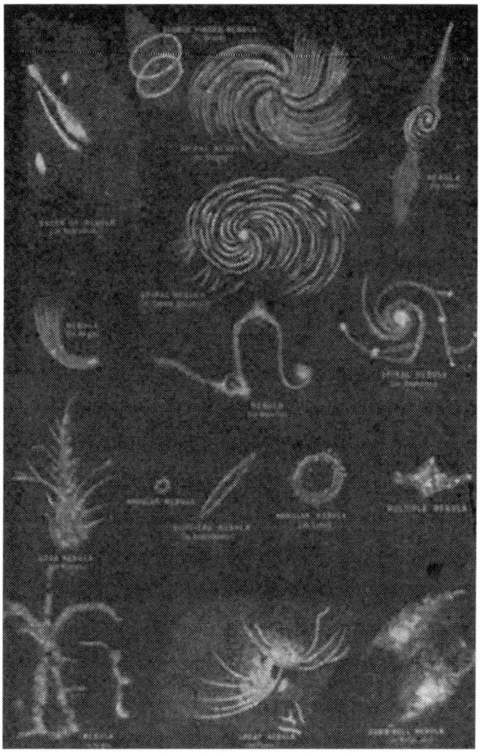

FIG. 141.— Types of nebulæ.

214. ***Nebulæ.***— Returning to Fig. 136, we note that its background has a hazy appearance, and that at its center the stars can no longer be distinguished, but blend one with another so as to appear like a bright cloud. The outer part of the cluster is *resolved* into stars, while in the picture the inner portion is not so resolved, although in the original photographic plate the individual stars can be distinguished to the very center of the cluster. In many cases, however, this is not possible, and we have an *irresolvable cluster* which it is customary to call a *nebula* (Latin, *little cloud*).

The most conspicuous example of this in the northern heavens is the great nebula in Andromeda (R. A. 0^h 37^m, Dec. +41°), which may be seen with the naked eye as a faint patch of foggy light. Look for it. This appears in an opera glass or very small telescope not unlike Fig. 137, which is reproduced from a sketch. Fig. 138 is from a photograph of the same object showing essentially the same shape as in the preceding figure, but bringing out more detail. Note the two small nebulæ adjoining the large one, and at the bottom of the picture an object which might easily be taken for another nebula but which is in fact a tailless comet that chanced to be passing that part of the sky when the picture was taken. Fig. 139 is from another drawing of this nebula, although it is hardly to be recognized as a representation of the same thing; but its characteristic feature, the two dark streaks near the center of the picture, is justified in part by Fig. 140, which is from a photograph made with a large reflecting telescope.

A comparison of these several representations of the same thing will serve to illustrate the vagueness of its outlines, and how much the impressions to be derived from nebulæ depend upon the telescopes employed and upon the observer's own prepossessions. The differences among the pictures can not be due to any change in the nebula itself, for half a century ago it was sketched much as shown in the latest of them (Fig. 140).

FIG. 142.— The Trifid nebula.— KEELER.

215. ***Typical nebulæ.***— Some of the fantastic forms which nebulæ present in the telescope are shown on a small scale in Fig. 141, but in recent years astronomers have learned to place little reliance upon drawings such as these, which are now almost entirely supplanted by photographs made with long exposures in powerful telescopes. One of the most exquisite of these modern photographs is that of the Trifid nebula in Sagittarius (Fig. 142). Note especially the dark lanes that give to this nebula its name, Trifid, and

which run through its brightest parts, breaking it into seemingly independent sections. The area of the sky shown in this cut is about 15 per cent less than that covered by the full moon.

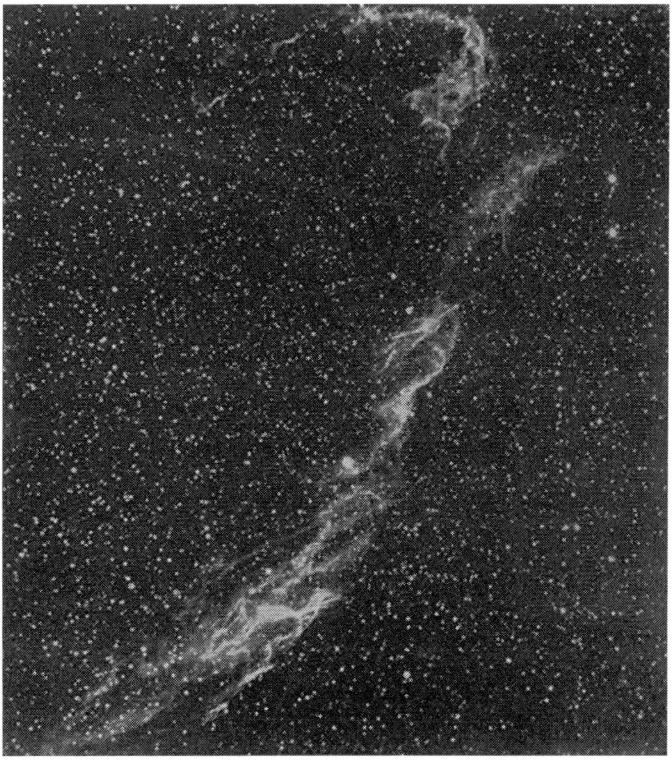

FIG. 143.—A nebula in Cygnus.—KEELER.

Fig. 143 shows a very different type of nebula, found in the constellation Cygnus, which appears made up of filaments closely intertwined, and stretches across the sky for a distance considerably greater than the moon's diameter.

FIG. 144.— Spiral nebula in Canes Venatici.— KEELER.

A much smaller but equally striking nebula is that in the constellation Canes Venatici (Fig. 144), which shows a most extraordinary spiral structure, as if the stars composing it were flowing in along curved lines toward a center of condensation. The diameter of the circular part of this nebula, omitting the projection toward the bottom of the picture, is about five minutes of arc, a sixth part of the diameter of the moon, and its thickness is probably very small compared with its breadth, perhaps not much exceeding the width of the spiral streams which compose it. Note how the bright stars that appear within the area of this nebula fall on the streams of nebulous matter as if they were part of them. This characteristic grouping of the stars, which is followed in many other nebulæ, shows that they are really part and parcel of the nebula and not merely on line with it. Fig. 145 shows how a great nebula is associated with the star ρ Ophiuchi.

FIG. 145.— Great nebula about the star ρ Ophiuchi.— BARNARD.

Probably the most impressive of all nebulæ is the great one in Orion (Fig. 146), whose position is shown on the star map between Rigel and ζ Orionis. Look for it with an opera glass or even with the unaided eye. This is sometimes called an *amorphous*— i. e., shapeless— nebula, because it presents no definite form which the eye can grasp and little trace of structure or organization. It is "without form and void" at least in its central portions, although on its edges curved filaments may be traced streaming away from the brighter parts of the central region. This nebula, as shown in Fig. 146, covers an area about equal to that of the full moon, without counting as any part of this the companion nebula shown at one side, but photographs made with suitable exposures show that faint outlying parts of the nebula extend in curved lines over the larger part of the constellation Orion. Indeed, over a large part of the entire sky the background is faintly covered with nebulous light whose brighter portions, if each were counted as a separate nebula, would carry the total number of such objects well into the hundreds of thousands.

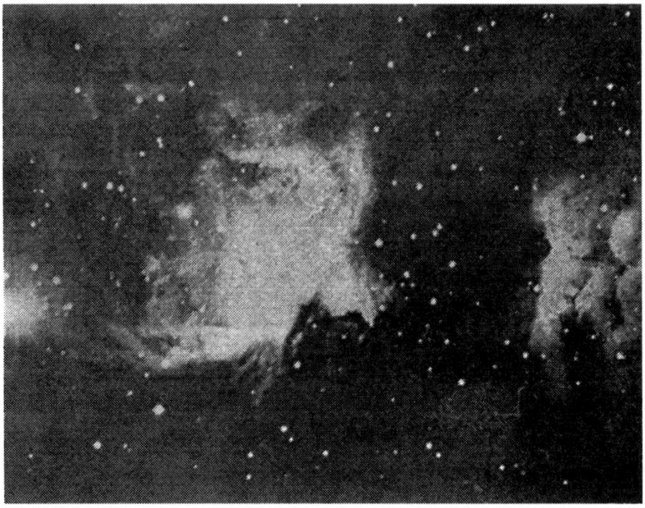

FIG. 146.— The Orion nebula.

The Pleiades (Plate IV) present a case of a resolvable star cluster projected against such a nebulous background whose varying intensity should be noted in the figure. A part of this nebulous matter is shown in wisps extending from one star to the next, after the fashion of a bridge, and leaving little doubt that the nebula is actually a part of the cluster and not merely a background for it.

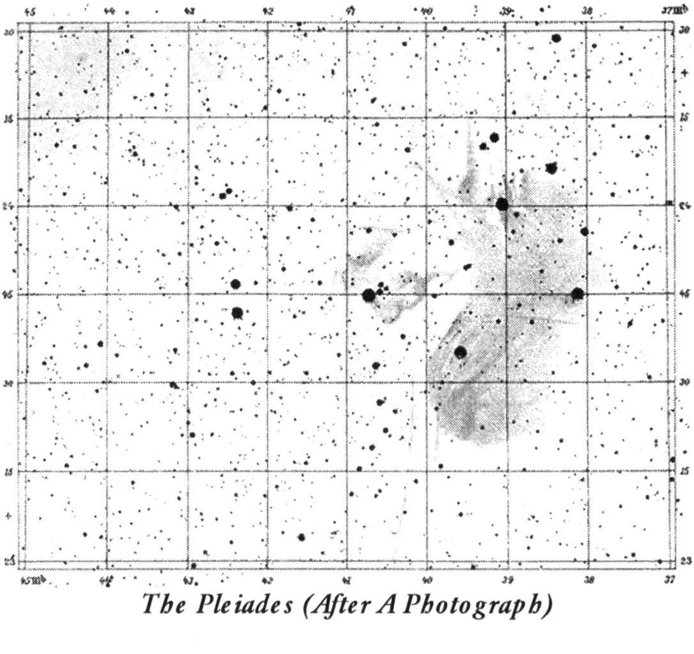

The Pleiades (After A Photograph)

Fig. 147 shows a series of so-called double nebulæ perhaps comparable with double stars, although the most recent photographic work seems to indicate that they are really faint spiral nebulæ in which only the brightest parts are shown by the telescope.

According to Keeler, the spiral is the prevailing type of nebulæ, and while Fig. 144 presents the most perfect example of such a nebula, the student should not fail to note that the Andromeda nebula (Fig. 140) shows distinct traces of a spiral structure, only here we do not see its true shape, the nebula being turned nearly edgewise toward us so that its presumably circular outline is foreshortened into a narrow ellipse.

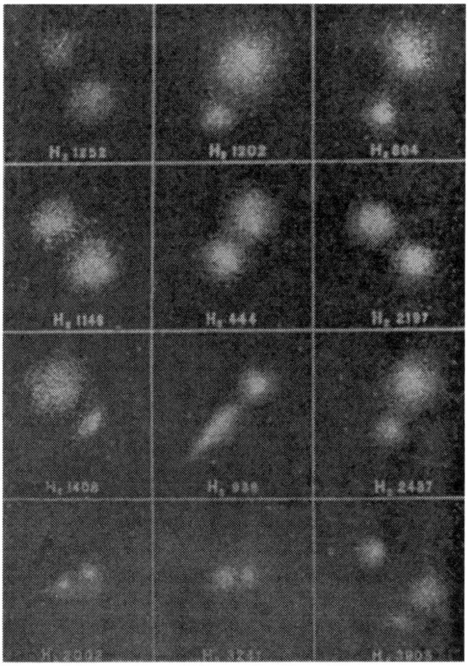

*FIG. 147.— **Double nebulæ.** HERSCHEL.*

Another type of nebula of some consequence presents in the telescope round disks like those of Uranus or Neptune, and this appearance has given them the name *planetary nebulæ*. The comet in Fig. 138, if smaller, would represent fairly well the nebulæ of this type. Sometimes a planetary nebula has a star at its center, and sometimes it appears hollow, like a smoke ring, and is then called a ring nebula. The most famous of these is in the constellation Lyra, not far from Vega.

216. ***Spectra of nebulæ.***— A star cluster, like the one in Hercules, shows, of course, stellar spectra, and even when irresolvable the spectrum is a

continuous one, testifying to the presence of stars, although they stand too close together to be separately seen. But in a certain number of nebulæ the spectrum is altogether different, a discontinuous one containing only a few bright lines, showing that here the nebular light comes from glowing gases which are subject to no considerable pressure. The planetary nebulæ all have spectra of this kind and make up about half of all the known gaseous nebulæ. It is worthy of note that a century ago Sir William Herschel had observed a green shimmer in the light of certain nebulæ which led him to believe that they were "not of a starry nature," a conclusion which has been abundantly confirmed by the spectroscope. The green shimmer is, in fact, caused by a line in the green part of the spectrum that is always present and is always the brightest part of the spectrum of gaseous nebulæ.

In faint nebulæ this line constitutes the whole of their visible spectrum, but in brighter ones two or three other and fainter lines are usually associated with it, and a very bright nebula, like that in Orion, may show a considerable number of extra lines, but for the most part they can not be identified in the spectrum of any terrestrial substances. An exception to this is found in the hydrogen lines, which are well marked in most spectra of gaseous nebulæ, and there are indications of one or two other known substances.

217. ***Density of nebulæ.***— It is known from laboratory experiments that diminishing the pressure to which an incandescent gas is subject, diminishes the number of lines contained in its spectrum, and we may surmise from the very simple character and few lines of these nebular spectra that the gas which produces them has a very small density. But this is far from showing that the nebula itself is correspondingly attenuated, for we must not assume that this shining gas is all that exists in the nebula; so far as telescope or camera are concerned, there may be associated with it any amount of dark matter which can not be seen because it sends to us no light. It is easy to think in this connection of meteoric dust or the stuff of which comets are made, for these seem to be scattered broadcast on every side of the solar system and may, perchance, extend out to the region of the nebulæ.

But, whatever may be associated in the nebula with the glowing gas which we see, the total amount of matter, invisible as well as visible, must be very small, or rather its average density must be very small, for the space occupied by such a nebula as that of Orion is so great that if the average density of its matter were equal to that of air the resulting mass by its attraction would exert a sensible effect upon the motion of the sun through space. The brighter parts of this nebula as seen from the earth subtend an angle of about half a degree, and while we know nothing of its distance from us, it is easy to see that the farther it is away the greater must be its real dimensions, and that this increase of bulk and mass with increasing distance will just compensate the diminishing intensity of gravity at great distances, so that for a given angular diameter— e. g., half a degree— the force with which this nebula attracts the sun depends upon its density but not at all upon its distance. Now, the nebula must attract the sun in some degree, and must tend to move it and the planets in an orbit about the attracting center so that year after year we should see the nebula from slightly different points of view, and this changed point of view should produce a change in the apparent direction of the nebula from us— i. e., a proper motion, whose amount would depend upon the attracting force, and therefore upon the density of the attracting matter. Observations of the Orion nebula show that its proper motion is wholly inappreciable, certainly far less than half a second of arc per year, and corresponding to this amount of proper motion the mean density of the nebula must be some millions of times (10^{10} according to Ranyard) less than that of air at sea level— i. e., the average density throughout the nebula is comparable with that of those upper parts of the earth's atmosphere in which meteors first become visible.

218. ***Motion of nebulæ.***— The extreme minuteness of their proper motions is a characteristic feature of all nebulæ. Indeed, there is hardly a known case of sensible proper motion of one of these bodies, although a dozen or more of them show velocities in the line of sight ranging in amount from +30 to -40 miles per second, the plus sign indicating an increasing distance. While a part of these velocities may be only apparent and due to the motion of earth and sun through space, a part at least is real motion of the nebulæ themselves. These seem to move through the celestial spaces in much the same way and with the same velocities as do the stars, and their smaller proper motions across the line of sight (angular motions) are an index of their great distance from us. No one has ever succeeded in measuring the parallax of a nebula or star cluster.

FIG. 148.—A part of the Milky Way.

The law of gravitation presumably holds sway within these bodies, and the fact that their several parts and the stars which are involved within them, although attracted by each other, have shown little or no change of position during the past century, is further evidence of their low density and feeble attraction. In a few cases, however, there seem to be in progress within a nebula changes of brightness, so that what was formerly a faint part has become a brighter one, or *vice versa*; but, on the whole, even these changes are very small.

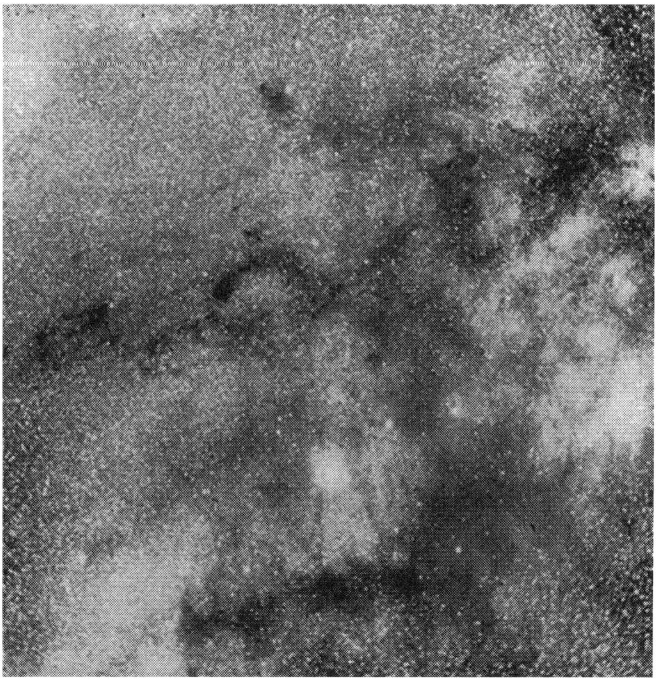

FIG. 149.— The Milky Way near θ Ophiuchi.— BARNARD.

219. ***The Milky Way.***— Closely related to nebulæ and star clusters is another feature of the sky, the *galaxy* or *Milky Way*, with whose appearance to the unaided eye the student should become familiar by direct study of the thing itself. Figs. 148 and 149 are from photographs of two small parts of it, and serve to bring out the small stars of which it is composed. Every star shown in these pictures is invisible to the naked eye, although their combined light is easily seen. The general course of the galaxy across the heavens is shown in the star maps, but these contain no indication of the wealth of detail which even the naked eye may detect in it. Bright and faint parts, dark rifts which cut it into segments, here and there a hole as if the ribbon of light had been shot away— such are some of the features to be found by attentive examination.

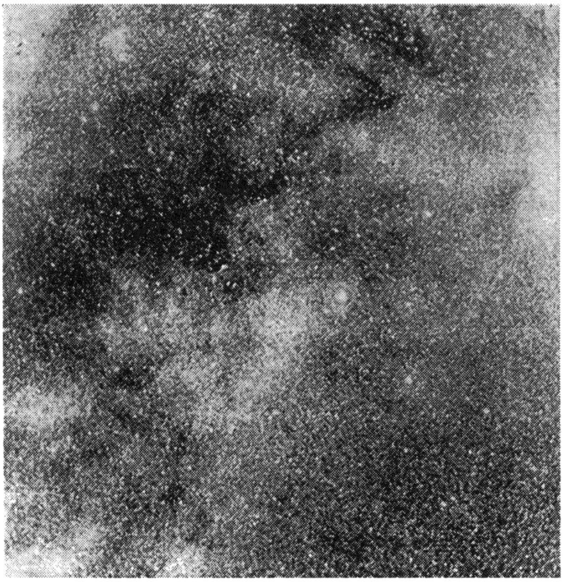

FIG. 150.— The Milky Way near β Cygni.— BARNARD.

Speaking generally, the course of the Milky Way is a great circle completely girdling the sky and having its north pole in the constellation Coma Berenices. The width of this stream of light is very different in different parts of the heavens, amounting where it is widest, in Lyra and Cygnus, to something more than 30°, although its boundaries are too vague and ill defined to permit much accuracy of measurement. Observe the very bright part between β and γ Cygni, nearly opposite Vega, and note how even an opera glass will partially resolve the nebulous light into a great number of stars, which are here rather brighter than in other parts of its course. But the resolution into stars is only partial, and there still remains a background of unresolved shimmer. Fig. 150 is a photograph of a small part of this region in which, although each fleck of light represents a separate star, the galaxy is not completely resolved. Compare with this region, rich in stars, the nearly empty space between the branches of the galaxy a little west of Altair. Another hole in the Milky Way may be found a little north and east of α Cygni, and between the extremes of abundance and poverty here noted there may be found every gradation of nebulous light.

The Milky Way is not so simple in its structure as might at first be thought, but a clear and moonless night is required to bring out its details. The nature of these details, the structure of the galaxy, its shape and extent, the arrangement of its parts, and their relation to stars and nebulæ in general, have been subjects of much speculation by astronomers and others who have sought to trace out in this way what is called the *construction of the heavens.*

220. ***Distribution of the stars.***– How far out into space do the stars extend? Are they limited or infinite in number? Do they form a system of mutually related parts, or are they bunched promiscuously, each for itself, without reference to the others? Here is what has been well called "the most important problem of stellar astronomy, the acquisition of well-founded ideas about the distribution of the stars." While many of the ideas upon this subject which have been advanced by eminent astronomers and which are still current in the books are certainly wrong, and few of their speculations along this line are demonstrably true, the theme itself is of such grandeur and permanent interest as to demand at least a brief consideration. But before proceeding to its speculative side we need to collect facts upon which to build, and these, however inadequate, are in the main simple and not far to seek.

Parallaxes, proper motions, motions in the line of sight, while pertinent to the problem of stellar distribution, are of small avail, since they are far too scanty in number and relate only to limited classes of stars, usually the very bright ones or those nearest to the sun. Almost the sole available data are contained in the brightness of the stars and the way in which they seem scattered in the sky. The most casual survey of the heavens is enough to show that the stars are not evenly sprinkled upon it. The lucid stars are abundant in some regions, few in others, and the laborious star gauges, actual counting of the stars in sample regions of the sky, which have been made by the Herschels, Celoria, and others, suffice to show that this lack of uniformity in distribution is even more markedly true of the telescopic stars.

The rate of increase in the number of stars from one magnitude to the next, as shown in § 187, is proof of another kind of irregularity in their distribution. It is not difficult to show, mathematically, that if in distant regions of space the stars were on the average as numerous and as bright as they are in the regions nearer to the sun, then the stars of any particular magnitude ought to be four times as numerous as those of the next brighter magnitude– e. g., four times as many sixth-magnitude stars as there are fifth-magnitude ones. But, as we have already seen in § 187, by actual count there are only three times as many, and from the discrepancy between these numbers, an actual threefold increase instead of a fourfold one, we must conclude that on the whole the stars near the sun are either bigger or brighter or more numerous than in the remoter depths of space.

221. ***The stellar system.***– But the arrangement of the stars is not altogether lawless and chaotic; there are traces of order and system, and among these the Milky Way is the dominant feature. Telescope and photographic plate alike show that it is made up of stars which, although quite irregularly scattered along its course, are on the average some twenty times as numerous in the galaxy as at its poles, and which thin out as we recede from it on either side, at first rapidly and then more slowly. This tendency to cluster

along the Milky Way is much more pronounced among the very faint telescopic stars than among the brighter ones, for the lucid stars and the telescopic ones down to the tenth or eleventh magnitude, while very plainly showing the clustering tendency, are not more than three times as numerous in the galaxy as in the constellations most remote from it. It is remarkable as showing the condensation of the brightest stars that one half of all the stars in the sky which are brighter than the second magnitude are included within a belt extending 12° on either side of the center line of the galaxy.

In addition to this general condensation of stars toward the Milky Way, there are peculiarities in the distribution of certain classes of stars which are worth attention. Planetary nebulæ and new stars are seldom, if ever, found far from the Milky Way, and stars with bright lines in their spectra especially affect this region of the sky. Stars with spectra of the first type– Sirian stars– are much more strongly condensed toward the Milky Way than are stars of the solar type, and in consequence of this the Milky Way is peculiarly rich in light of short wave lengths. Resolvable star clusters are so much more numerous in the galaxy than elsewhere, that its course across the sky would be plainly indicated by their grouping upon a map showing nothing but clusters of this kind.

On the other hand, nebulæ as a class show a distinct aversion for the galaxy, and are found most abundantly in those parts of the sky farthest from it, much as if they represented raw material which was lacking along the Milky Way, because already worked up to make the stars which are there so numerous.

222. ***Relation of the sun to the Milky Way.***– The fact that the galaxy is a *great circle* of the sky, but only of moderate width, shows that it is a widely extended and comparatively thin stratum of stars within which the solar system lies, a member of the galactic system, and probably not very far from its center. This position, however, is not to be looked upon as a permanent one, since the sun's motion, which lies nearly in the plane of the Milky Way, is ceaselessly altering its relation to the center of that system, and may ultimately carry us outside its limits.

The Milky Way itself is commonly thought to be a ring, or series of rings, like the coils of the great spiral nebula in Andromeda, and separated from us by a space far greater than the thickness of the ring itself. Note in Figs. 149 and 150 how the background is made up of bright and dark parts curiously interlaced, and presenting much the appearance of a thin sheet of cloud through which we look to barren space beyond. While, mathematically, this appearance can not be considered as proof that the galaxy is in fact a distant ring, rather than a sheet of starry matter stretching continuously from the nearer stellar neighbors of the sun into the remotest depths of space, nevertheless, most students of the question hold it to be such a ring of stars, which are relatively close together while its center is comparatively vacant, although even here are some hundreds of thousands of stars which on the whole have a tendency to cluster near its plane and to crowd together a little more densely than elsewhere in the region where the sun is placed.

223. ***Dimensions of the galaxy.***— The dimensions of this stellar system are wholly unknown, but there can be no doubt that it extends farther in the plane of the Milky Way than at right angles to that plane, for stars of the fifteenth and sixteenth magnitudes are common in the galaxy, and testify by their feeble light to their great distance from the earth, while near the poles of the Milky Way there seem to be few stars fainter than the twelfth magnitude. Herschel, with his telescope of 18 inches aperture, could count in the Milky Way more than a dozen times as many stars per square degree as could Celoria with a telescope of 4 inches aperture; but around the poles of the galaxy the two telescopes showed practically the same number of stars, indicating that here even the smaller telescope reached to the limits of the stellar system. Very recently, indeed, the telescope with which Fig. 140 was photographed seems to have reached the farthest limit of the Milky Way, for on a photographic plate of one of its richest regions Roberts finds it completely resolved into stars which stand out upon a black background with no trace of nebulous light between them.

224. ***Beyond the Milky Way.***— Each additional step into the depths of space brings us into a region of which less is known, and what lies beyond the Milky Way is largely a matter of conjecture. We shrink from thinking it an infinite void, endless emptiness, and our intellectual sympathies go out to Lambert's speculation of a universe filled with stellar systems, of which ours, bounded by the galaxy, is only one. There is, indeed, little direct evidence that other such systems exist, but the Andromeda nebula is not altogether unlike a galaxy with a central cloud of stars, and in the southern hemisphere, invisible in our latitudes, are two remarkable stellar bodies like the Milky Way in appearance, but cut off from all apparent connection with it, much as we might expect to find independent stellar systems, if such there be.

These two bodies are known as the Magellanic clouds, and individually bear the names of Major and Minor Nubecula. According to Sir John Herschel, "the Nubecula Major, like the Minor, consists partly of large tracts and ill-defined patches of irresolvable nebula, and of nebulosity in every stage of resolution up to perfectly resolved stars like the Milky Way, as also of regular and irregular nebulæ ... of globular clusters in every stage of resolvability, and of clustering groups sufficiently insulated and condensed to come under the designation of clusters of stars." Its outlines are vague and somewhat uncertain, but surely include an area of more than 40 square degrees— i. e., as much as the bowl of the Big Dipper— and within this area Herschel counted several hundred nebulæ and clusters "which far exceeds anything that is to be met with in any other region of the heavens." Although its excessive complexity of detail baffled Herschel's attempts at artistic delineation, it has yielded to the modern photographic processes, which show the Nubecula Major to be an enormous spiral nebula made up of subordinate stars, nebulæ, and clusters, as is the Milky Way.

Compared with the Andromeda nebula, its greater angular extent suggests a smaller distance, although for the present all efforts at determining the parallax of either seem hopeless. But the spiral form which is common to both suggests that the Milky Way itself may be a gigantic spiral nebula near whose center lies the sun, a humble member of a great cluster of stars which is roughly globular in shape, but flattened at the poles of the galaxy and completely encircled by its coils. However plausible such a view may appear, it is for the present, at least, pure hypothesis, although vigorously advocated by Easton, who bases his argument upon the appearance of the galaxy itself.

225. ***Absorption of starlight.*** – We have had abundant occasion to learn that at least within the confines of the solar system meteoric matter, cosmic dust, is profusely scattered, and it appears not improbable that the same is true, although in smaller degree, in even the remoter parts of space. In this case the light which comes from the farther stars over a path requiring many centuries to travel, must be in some measure absorbed and enfeebled by the obstacles which it encounters on the way. Unless celestial space is transparent to an improbable degree the remoter stars do not show their true brightness; there is a certain limit beyond which no star is able to send its light, and beyond which the universe must be to us a blank. A lighthouse throws into the fog its beams only to have them extinguished before a single mile is passed, and though the celestial lights shine farther, a limit to their reach is none the less certain if meteoric dust exists outside the solar system. If there is such an absorption of light in space, as seems plausible, the universe may well be limitless and the number of stellar systems infinite, although the most attenuated of dust clouds suffices to conceal from us and to shut off from our investigation all save a minor fraction of it and them.

CHAPTER XV.
GROWTH AND DECAY

226. ***Nature of the problem.***– To use a common figure of speech, the universe is alive. We have found it filled with an activity that manifests itself not only in the motions of the heavenly bodies along their orbits, but which extends to their minutest parts, the molecules and atoms, whose vibrations furnish the radiant energy given off by sun and stars. Some of these activities, such as the motions of the heavenly bodies in their orbits, seem fitted to be of endless duration; while others, like the radiation of light and heat, are surely temporary, and sooner or later must come to an end and be replaced by something different. The study of things as they are thus leads inevitably to questions of what has been and what is to be. A sound science should furnish some account of the universe of yesterday and to-morrow as well as of to-day, and we need not shrink from such questions, although answers to them must be vague and in great measure speculative.

The historian of America finds little difficulty with events of the nineteenth century or even the eighteenth, but the sources of information about America in the fifteenth century are much less definite; the tenth century presents almost a blank, and the history of American mankind in the first century of the Christian era is wholly unknown. So, as we attempt to look into the past or the future of the heavens, we must expect to find the mists of obscurity grow denser with remoter periods until even the vaguest outlines of its development are lost, and we are compelled to say, beyond this lies the unknown. Our account of growth and decay in the universe, therefore, can not aspire to cover the whole duration of things, but must be limited in its scope to certain chapters whose epochs lie near to the time in which we live, and even for these we need to bear constantly in mind the logical bases of such an inquiry and the limitations which they impose upon us.

227. ***Logical bases and limitations.***– The first of these bases is: An adequate knowledge of the present universe. Our only hope of reading the past and future lies in an understanding of the present; not necessarily a complete knowledge of it, but one which is sound so far as it goes. Our position is like that of a detective who is called upon to unravel a mystery or crime, and who must commence with the traces that have been left behind in its commission. The foot print, the blood stain, the broken glass must be examined and compared, and fashioned into a theory of how they came to be; and as a wrong understanding of these elements is sure to vitiate the theories based upon them, so a false science of the universe as it now is, will surely give a false account of what it has been; while a correct but incomplete knowledge of the present does not wholly bar an understanding of the past,

but only puts us in the position of the detective who correctly understands what he sees but fails to take note of other facts which might greatly aid him.

The second basis of our inquiry is: The assumed permanence of natural laws. The law of gravitation certainly held true a century ago as well as a year ago, and for aught we know to the contrary it may have been a law of the universe for untold millions of years; but that it has prevailed for so long a time is a pure assumption, although a necessary one for our purpose. So with those other laws of mathematics and mechanics and physics and chemistry to which we must appeal; if there was ever a time or place in which they did not hold true, that time and place lie beyond the scope of our inquiry, and are in the domain inaccessible to scientific research. It is for this reason that science knows nothing and can know nothing of a creation or an end of the universe, but considers only its orderly development within limited periods of time. What kind of a past universe would, under the operation of known laws, develop into the present one, is the question with which we have to deal, and of it we may say with Helmholtz: "From the standpoint of science this is no idle speculation but an inquiry concerning the limitations of its methods and the scope of its known laws."

To ferret out the processes by which the heavenly bodies have been brought to their present condition we seek first of all for lines of development now in progress which tend to change the existing order of things into something different, and, having found these, to trace their effects into both past and future. Any force, however small, or any process, however slow, may produce great results if it works always and ceaselessly in the same direction, and it is in these processes, whose trend is never reversed, that we find a partial clew to both past and future.

228. ***The sun's development.***— The first of these to claim our attention is the shrinking of the sun's diameter which, as we have seen in Chapter X, is the means by which the solar output of radiant energy is maintained from year to year. Its amount, only a few feet per annum, is far too small to be measured with any telescope; but it is cumulative, working century after century in the same direction, and, given time enough, it will produce in the future, and must have produced in the past, enormous transformations in the sun's bulk and equally significant changes in its physical condition.

Thus, as we attempt to trace the sun's history into the past, the farther back we go the greater shall we expect to find its diameter and the greater the space (volume) through which its molecules are spread. By reason of this expansion its density must have been less then than now, and by going far enough back we may even reach a time at which the density was comparable with what we find in the nebulæ of to-day. If our ideas of the sun's present mechanism are sound, then, as a necessary consequence of these, its past career must have been a process of condensation in which its component particles were year by year packed closer together by their own attraction for each other. As we have seen in § 126, this condensation necessarily developed heat, a part of which was radiated away as fast as produced, while the remainder was stored up, and served to raise the temperature of the sun to what we find it now. At the present time this temperature is a chief obstacle to further shrinkage, and so powerfully opposes the gravitative forces as to maintain nearly an equilibrium with them, thus causing a very slow rate of further condensation. But it is not probable that this was always so. In the early stages of the sun's history, when the temperature was low, contraction of its bulk must have been more rapid, and attempts have been made by the mathematicians to measure its rate of progress and to determine how long a time has been consumed in the development of the present sun from a primitive nebulous condition in which it filled a space of greater diameter than Neptune's orbit. Of course, numerical precision is not to be expected in results of this kind, but, from a consideration of the greatest amount of heat that could be furnished by the shrinkage of a mass equal to that of the sun, it seems that the period of this development is to be measured in tens of millions or possibly hundreds of millions of years, but almost certainly does not reach a thousand millions.

229. ***The sun's future.***— The future duration of the sun as a source of radiant energy is surely to be measured in far smaller numbers than these. Its career as a dispenser of light and heat is much more than half spent, for the shrinkage results in an ever-increasing density, which makes its gaseous substance approximate more and more toward the behavior of a liquid or solid, and we recall that these forms of matter can not by any further condensation restore the heat whose loss through radiation caused them to contract. They may continue to shrink, but their temperature must fall, and when the sun's substance becomes too dense to obey the laws of gaseous matter its surface must cool rapidly as a consequence of the radiation into surrounding space, and must congeal into a crust which, although at first incandescent, will speedily become dark and opaque, cutting off the light of the central portions, save as it may be rent from time to time by volcanic outbursts of the still incandescent mass beneath. But such outbursts can be of short duration only, and its final condition must be that of a dark body, like the earth or moon, no longer available as a source of radiant energy. Even before the formation of a solid crust it is quite possible that the output of light and heat may be seriously diminished by the formation of dense vapors completely enshrouding it, as is now the case with Jupiter and Saturn. It is believed that these planets were formerly incandescent, and at the present time are in a state of development through which the earth has passed and toward which the sun is moving. According to Newcomb, the future during which the sun can continue to furnish light and heat at its present rate is not likely to exceed 10,000,000 years.

This idea of the sun as a developing body whose present state is only temporary, furnishes a clew to some of the vexing problems of solar physics. Thus the sun-spot period, the distribution of the spots in latitude, and the peculiar law of rotation of the sun in different latitudes, may be, and very probably are, results not of anything now operating beneath its photosphere, but of something which happened to it in the remote past– e. g., an unsymmetrical shrinkage or possibly a collision with some other body. At sea the waves continue to toss long after the storm which produced them has disappeared, and, according to the mathematical researches of Wilsing, a profound agitation of the sun's mass might well require tens of thousands, or even hundreds of thousands of years to subside, and during this time its effects would be visible, like the waves, as phenomena for which the actual condition of things furnishes no apparent cause.

230. ***The nebular hypothesis.*** – The theory of the sun's progressive contraction as a necessary result of its radiation of energy is comparatively modern, but more than a century ago philosophic students of Nature had been led in quite a different way to the belief that in the earlier stages of its career the sun must have been an enormously extended body whose outer portions reached even beyond the orbit of the remotest planet. Laplace, whose speculations upon this subject have had a dominant influence during the nineteenth century, has left, in a popular treatise upon astronomy, an admirable statement of the phenomena of planetary motion, which suggest and lead up to the nebular theory of the sun's development, and in presenting this theory we shall follow substantially his line of thought, but with some freedom of translation and many omissions.

He says: "To trace out the primitive source of the planetary movements, we have the following five phenomena: (1) These movements all take place in the same direction and nearly in the same plane. (2) The movements of the satellites are in the same direction as those of the planets. (3) The rotations of the planets and the sun are in the same direction as the orbital motions and nearly in the same plane. (4) Planets and satellites alike have nearly circular orbits. (5) The orbits of comets are wholly unlike these by reason of their great eccentricities and inclinations to the ecliptic." That these coincidences should be purely the result of chance seemed to Laplace incredible, and, seeking a cause for them, he continues: "Whatever its nature may be, since it has produced or controlled the motions of the planets, it must have reached out to all these bodies, and, in view of the prodigious distances which separate them, the cause can have been nothing else than a fluid of great extent which must have enveloped the sun like an atmosphere. A consideration of the planetary motions leads us to think that ... the sun's atmosphere formerly extended far beyond the orbits of all the planets and has shrunk by degrees to its present dimensions." This is not very different from the idea developed in § 228 from a consideration of the sun's radiant energy; but in Laplace's day the possibility of generating the sun's heat by contraction of its bulk was unknown, and he was compelled to assume a very high temperature for the primitive nebulous sun, while we now know that this is unnecessary. Whether the primitive nebula was hot or cold the shrinkage would take place in much the same way, and would finally result in a star or sun of very high temperature, but its development would be slower if it were hot in the beginning than if it were cold.

But again Laplace: "How did the sun's atmosphere determine the rotations and revolutions of planets and satellites? If these bodies had been deeply immersed in this atmosphere its resistance to their motion would have made them fall into the sun, and we may therefore conjecture that the planets were formed, one by one, at the outer limits of the solar atmosphere by the condensation of zones of vapor which were cast off in the plane of the sun's equator." Here he proceeds to show by an appeal to dynamical principles that something of this kind must happen, and that the matter sloughed off by the nebula in the form of a ring, perhaps comparable to the rings of Saturn or the asteroid zone, would ultimately condense into a planet, which in its turn might shrink and cast off rings to produce satellites.

Pierre Simon Laplace (1749-1827)

Planets and satellites would then all have similar motions, as noted at the beginning of this section, since in every case this motion is an inheritance from a common source, the rotation of the primitive nebula about its own axis. "All the bodies which circle around a planet having been thus formed from rings which its atmosphere successively abandoned as rotation became more and more rapid, this rotation should take place in less time than is required for the orbital revolution of any of the bodies which have been cast off, and this holds true for the sun as compared with the planets."

231. ***Objections to the nebular hypothesis.***— In Laplace's time this slower rate of motion was also supposed to hold true for Saturn's rings as compared with the rotation of Saturn itself, but, as we have seen in Chapter XI, this ring is made up of a great number of independent particles which move at different rates of speed, and comparing, through Kepler's Third Law, the motion of the inner edge of the ring with the known periodic time of the satellites, we may find that these particles must rotate about Saturn more rapidly than the planet turns upon its axis. Similarly the inner satellite of Mars completes its revolution in about one third of a Martian day, and we find in cases like this grounds for objection to the nebular theory. Compare also Laplace's argument with the peculiar rotations of Uranus, Neptune, and their satellites (Chapter XI). Do these fortify or weaken his case?

Despite these objections and others equally serious that have been raised, the nebular theory agrees with the facts of Nature at so many points that astronomers upon the whole are strongly inclined to accept its major outlines as being at least an approximation to the course of development actually followed by the solar system; but at some points— e. g., the formation of planets and satellites through the casting off of nebulous rings— the objections are so many and strong as to call for revision and possibly serious modification of the theory.

One proposed modification, much discussed in recent years, consists in substituting for the primitive *gaseous* nebula imagined by Laplace, a very diffuse cloud of meteoric matter which in the course of its development would become transformed into the gaseous state by rising temperature. From this point of view much of the meteoric dust still scattered throughout the solar system may be only the fragments left over in fashioning the sun and planets. Chamberlin and Moulton, who have recently given much attention to this subject, in dissenting from some of Laplace's views, consider that the primitive nebulous condition must have been one in which the matter of the system was "so brought together as to give low mass, high momentum, and irregular distribution to the outer part, and high mass, low momentum, and sphericity to the central part," and they suggest a possible oblique collision of a small nebula with the outer parts of a large one.

232. ***Bode's law.***— We should not leave the theory of Laplace without noting the light it casts upon one point otherwise obscure— the meaning of Bode's law (§ 134). This law, stated in mathematical form, makes a geometrical series, and similar geometrical series apply to the distances of the

satellites of Jupiter and Saturn from these planets. Now, Roche has shown by the application of physical laws to the shrinkage of a gaseous body that its radius at any time may be expressed by means of a certain mathematical formula very similar to Bode's law, save that it involves the amount of time that has elapsed since the beginning of the shrinking process. By comparing this formula with the one corresponding to Bode's law he reaches the conclusion that the peculiar spacing of the planets expressed by that law means that they were formed at successive *equal* intervals of time– i. e., that Mars is as much older than the earth as the earth is older than Venus, etc. The failure of Bode's law in the case of Neptune would then imply that the interval of time between the formation of Neptune and Uranus was shorter than that which has prevailed for the other planets. But too much stress should not be placed upon this conclusion. So long as the manner in which the planets came into being continues an open question, conclusions about their time of birth must remain of doubtful validity.

233. ***Tidal friction between earth and moon.***– An important addition to theories of development within the solar system has been worked out by Prof. G. H. Darwin, who, starting with certain very simple assumptions as to the present condition of things in earth and moon, derives from these, by a strict process of mathematical reasoning, far-reaching conclusions of great interest and importance. The key to these conclusions lies in recognition of the fact that through the influence of the tides (§ 42) there is now in progress and has been in progress for a very long time, a gradual transfer of motion (moment of momentum) from the earth to the moon. The earth's motion of rotation is being slowly destroyed by the friction of the tides, as the motion of a bicycle is destroyed by the friction of a brake, and, in consequence of this slowing down, the moon is pushed farther and farther away from the earth, so that it now moves in a larger orbit than it had some millions of years ago.

Fig. 24 has been used to illustrate the action of the moon in raising tides upon the earth, but in accordance with the third law of motion (§ 36) this action must be accompanied by an equal and contrary reaction whose nature may readily be seen from the same figure. The moon moves about its orbit from west to east and the earth rotates about its axis in the same direction, as shown by the curved arrow in the figure. The tidal wave, *I*, therefore points a little *in advance* of the moon's position in its orbit and by its attraction must tend to pull the moon ahead in its orbital motion a little faster than it would move if the whole substance of the earth were placed inside the sphere represented by the broken circle in the figure. It is true that the tidal wave at *I''* pulls back and tends to neutralize the effect of the wave at *I*, but on the whole the tidal wave nearer the moon has the stronger influence, and the moon on the whole moves a very little faster, and by virtue of this added impetus draws continually a little farther away from the earth than it would if there were no tides.

234. ***Consequences of tidal friction upon the earth.***— This process of moving the moon away from the earth is a cumulative one, going on century after century, and with reference to it the moon's orbit must be described not as a circle or ellipse, or any other curve which returns into itself, but as a spiral, like the balance spring of a watch, each of whose coils is a little larger than the preceding one, although this excess is, to be sure, very small, because the tides themselves are small and the tidal influence feeble when compared with the whole attraction of the earth for the moon. But, given time enough, even this small force may accomplish great results, and something like 100,000,000 years of past opportunity would have sufficed for the tidal forces to move the moon from close proximity with the earth out to its present position.

For millions of years to come, if moon and earth endure so long, the distance between them must go on increasing, although at an ever slower rate, since the farther away the moon goes the smaller will be the tides and the slower the working out of their results. On the other hand, when the moon was nearer the earth than now, tidal influences must have been greater and their effects more rapidly produced than at the present time, particularly if, as seems probable, at some past epoch the earth was hot and plastic like Jupiter and Saturn. Then, instead of tides in the water of the sea, such as we now have, the whole substance of the earth would respond to the moon's attraction in *bodily tides* of semi-fluid matter not only higher, but with greater internal friction of their molecules one upon another, and correspondingly greater effect in checking the earth's rotation.

But, whether the tide be a bodily one or confined to the waters of the sea, so long as the moon causes it to flow there will be a certain amount of friction which will affect the earth much as a brake affects a revolving wheel, slowing down its motion, and producing thus a longer day as well as a longer month on account of the moon's increased distance. Slowing down the earth's rotation is the direct action of the moon upon the earth. Pushing the moon away is the form in which the earth's equal and contrary reaction manifests itself.

235. ***Consequences of tidal friction upon the moon.***— When the moon was plastic the earth must have raised in it a bodily tide manifold greater than the lunar tides upon the earth, and, as we have seen in Chapter IX, this tide has long since worn out the greater part of the moon's rotation and brought our satellite to the condition in which it presents always the same face toward the earth.

These two processes, slowing down the rotation and pushing away the disturbing body, are inseparable— one requires the other; and it is worth noting in this connection that when for any reason the tide ceases to flow, and the tidal wave takes up a permanent position, as it has in the moon (§ 99), its work is ended, for when there is no motion of the wave there can be no friction to further reduce the rate of rotation of the one body, and no reaction to that friction to push away the other. But this permanent and stationary tidal wave in the moon, or elsewhere, means that the satellite presents always the same face toward its planet, moving once about its orbit in the time required for one revolution upon its axis, and the tide raised by the moon upon the earth tends to produce here the result long since achieved in our satellite, to make our day and month of equal length, and to make the earth turn always the same side toward the moon. But the moon's tidal force is small compared with that of the earth, and has a vastly greater momentum to overcome, so that its work upon the earth is not yet complete. According to Thomson and Tait, the moon must be pushed off another hundred thousand miles, and the day lengthened out by tidal influence to seven of our present weeks before the day and the lunar month are made of equal length, and the moon thereby permanently hidden from one hemisphere of the earth.

236. ***The earth-moon system.*** – Retracing into the past the course of development of the earth and moon, it is possible to reach back by means of the mathematical theory of tidal friction to a time at which these bodies were much nearer to each other than now, but it has not been found possible to trace out the mode of their separation from one body into two, as is supposed in the nebular theory. In the earliest part of their history accessible to mathematical analysis they are distinct bodies at some considerable distance from each other, with the earth rotating about an axis more nearly perpendicular to the moon's orbit and to the ecliptic than is now the case. Starting from such a condition, the lunar tides, according to Darwin, have been instrumental in tipping the earth's rotation axis into its present oblique position, and in determining the eccentricity of the moon's orbit and its position with respect to the ecliptic as well as the present length of day and month.

237. ***Tidal friction upon the planets.*** – The satellites of the outer planets are equally subject to influences of this kind, and there appears to be independent evidence that some of them, at least, turn always the same face toward their respective planets, indicating that the work of tidal friction has here been accomplished. We saw in Chapter XI that it is at present an open question whether the inner planets, Venus and Mercury, do not always turn the same face toward the sun, their day and year being of equal length. In addition to the direct observational evidence upon this point, Schiaparelli has sought to show by an appeal to tidal theory that such is probably the case, at least for Mercury, since the tidal forces which tend to bring about this result in that planet are about as great as the forces which have certainly produced it in the case of the moon and Saturn's satellite, Japetus. The same line of

reasoning would show that every satellite in the solar system, save possibly the newly discovered ninth satellite of Saturn, must, as a consequence of tidal friction, turn always the same face toward its planet.

238. ***The solar tide.***— The sun also raises tides in the earth, and their influence must be similar in character to that of the lunar tides, checking the rotation of the earth and thrusting earth and sun apart, although quantitatively these effects are small compared with those of the moon. They must, however, continue so long as the solar tide lasts, possibly until the day and year are made of equal length— i. e., they may continue long after the lunar tidal influence has ceased to push earth and moon apart. Should this be the case, a curious inverse effect will be produced. The day being then longer than the month, the moon will again raise a tide in the earth which will run around it *from west to east*, opposite to the course of the present tide, thus tending to accelerate the earth's rotation, and by its reaction to bring the moon back toward the earth again, and ultimately to fall upon it.

We may note that an effect of this kind must be in progress now between Mars and its inner satellite, Phobos, whose time of orbital revolution is only one third of a Martian day. It seems probable that this satellite is in the last stages of its existence as an independent body, and must ultimately fall into Mars.

239. ***Roche's limit.***— In looking forward to such a catastrophe, however, due regard must be paid to a dynamical principle of a different character. The moon can never be precipitated upon the earth entire, since before it reaches us it will have been torn asunder by the excess of the earth's attraction for the near side of its satellite over that which it exerts upon the far side. As the result of Roche's mathematical analysis we are able to assign a limiting distance between any planet and its satellite within which the satellite, if it turns always the same face toward the planet, can not come without being broken into fragments. If we represent the radius of the planet by r, and the quotient obtained by dividing the density of the planet by the density of the satellite by q, then

Roche's limit = $2.44\, r \sqrt[3]{q}$.

Thus in the case of earth and moon we find from the densities given in § 95, q = 1.65, and with r = 3,963 miles we obtain 11,400 miles as the nearest approach which the moon could make to the earth without being broken up by the difference of the earth's attractions for its opposite sides.

We must observe, however, that Roche's limit takes no account of molecular forces, the adhesion of one molecule to another, by virtue of which a stick or stone resists fracture, but is concerned only with the gravitative forces by which the molecules are attracted toward the moon's center and toward the earth. Within a stone or rock of moderate size these gravitative forces are insignificant, and cohesion is the chief factor in preserving its integrity, but in a large body like the moon, the case is just reversed, cohesion plays a small part and gravitation a large one in holding the body together. We may conclude, therefore, that at a proper distance these forces are capable of breaking up the moon, or any other large body, into fragments of a size such that molecular cohesion instead of gravitation is the chief agent in preserving them from further disintegration.

240. ***Saturn's rings.***— Saturn's rings are of peculiar interest in this connection. The outer edge of the ring system lies just inside of Roche's limit for this planet, and we have already seen that the rings are composed of small fragments independent of each other. Whatever may have been the process by which the nine satellites of Saturn came into existence, we have in Roche's limit the explanation why the material of the ring was not worked up into satellites; the forces exerted by Saturn would tear into pieces any considerable satellite thus formed and equally would prevent the formation of one from raw material.

Saturn's rings present the only case within the solar system where matter is known to be revolving about a planet at a distance less than Roche's limit, and it is an interesting question whether these rings can remain as a permanent part of the planet's system or are only a temporary feature. The drawings of Saturn made two centuries ago agree among themselves in representing the rings as larger than they now appear, and there is some reason to suppose that as a consequence of mutual disturbances— collisions— their momentum is being slowly wasted so that ultimately they must be precipitated into the planet. But the direct evidence of such a progress that can be drawn from present data is too scanty to justify positive conclusions in the matter. On the other hand, Nolan suggests that in the outer parts of the ring small satellites might be formed whose tidal influence upon Saturn would suffice to push them away from the ring beyond Roche's limit, and that the very small inner satellites of Saturn may have been thus formed at the expense of the ring.

The inner satellite of Mars is very close to Roche's limit for that planet, and, as we have seen above, must be approaching still nearer to the danger line.

241. ***The moon's development.***— The fine series of photographs of the moon obtained within the last few years at Paris, have been used by the astronomers of that observatory for a minute study of the lunar formations, much as geologists study the surface of the earth to determine something about the manner in which it was formed. Their conclusions are, in general, that at some past time the moon was a hot and fluid body which, as it cooled and condensed, formed a solid crust whose further shrinkage compressed the liquid nucleus and led to a long series of fractures in the crust and outbursts of liquid matter, whose latest and feeblest stages produced the lunar craters, while traces of the earlier ones, connected with a general settling of the crust,

although nearly obliterated, are still preserved in certain large but vague features of the lunar topography, such as the distribution of the seas, etc. They find also in certain markings of the surface what they consider convincing evidence of the existence in past times of a lunar atmosphere. But this seems doubtful, since the force of gravity at the moon's surface is so small that an atmosphere similar to that of the earth, even though placed upon the moon, could not permanently endure, but would be lost by the gradual escape of its molecules into the surrounding space.

The molecules of a gas are quite independent one of another, and are in a state of ceaseless agitation, each one darting to and fro, colliding with its neighbors or with whatever else opposes its forward motion, and traveling with velocities which, on the average, amount to a good many hundreds of feet per second, although in the case of any individual molecule they may be much less or much greater than the average value, an occasional molecule having possibly a velocity several times as great as the average. In the upper regions of our own atmosphere, if one of these swiftly moving particles of oxygen or nitrogen were headed away from the earth with a velocity of seven miles per second, the whole attractive power of the earth would be insufficient to check its motion, and it would therefore, unless stopped by some collision, escape from the earth and return no more. But, since this velocity of seven miles per second is more than thirty times as great as the average velocity of the molecules of air, it must be very seldom indeed that one is found to move so swiftly, and the loss of the earth's atmosphere by leakage of this sort is insignificant. But upon the moon, or any other body where the force of gravity is small, conditions are quite different, and in our satellite a velocity of little more than one mile per second would suffice to carry a molecule away from the outer limits of its atmosphere. This velocity, only five times the average, would be frequently attained, particularly in former times when the moon's temperature was high, for then the average velocity of all the molecules would be considerably increased, and the amount of leakage might become, and probably would become, a serious matter, steadily depleting the moon's atmosphere and leading finally to its present state of exhaustion. It is possible that the moon may at one time have had an atmosphere, but if so it could have been only a temporary possession, and the same line of reasoning may be applied to the asteroids and to most of the satellites of the solar system, and also, though in less degree, to the smaller planets, Mercury and Mars.

242. ***Stellar development.***— We have already considered in this chapter the line of development followed by one star, the sun, and treating this as a typical case, it is commonly believed that the life history of a star, in so far as it lies within our reach, begins with a condition in which its matter is widely diffused, and presumably at a low temperature. Contracting in bulk under the influence of its own gravitative forces, the star's temperature rises to a

maximum, and then falls off in later stages until the body ceases to shine and passes over to the list of dark stars whose existence can only be detected in exceptional cases, such as are noted in Chapter XIII. The most systematic development of this idea is due to Lockyer, who looks upon all the celestial bodies— sun, moon and planets, stars, nebulæ, and comets— as being only collections of meteoric matter in different stages of development, and who has sought by means of their spectra to classify these bodies and to determine their stage of advancement. While the fundamental ideas involved in this "meteoritic hypothesis" are not seriously controverted, the detailed application of its principles is open to more question, and for the most part those astronomers who hold that in the present state of knowledge stellar spectra furnish a key to a star's age or degree of advancement do not venture beyond broad general statements.

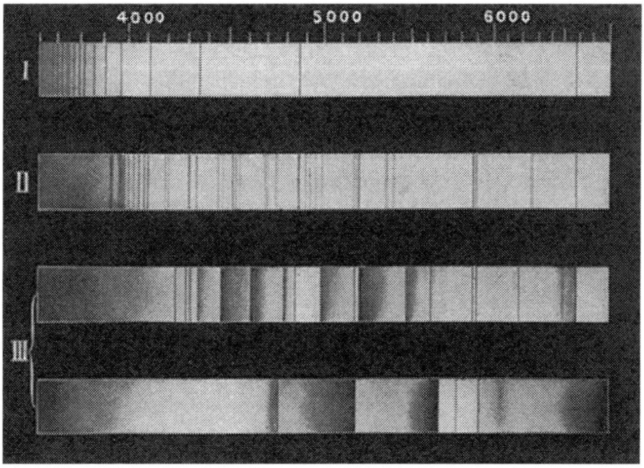

FIG. 151.— Types of stellar spectra substantially according to SECCHI.

243. ***Stellar spectra.***— Thus the types of stellar spectra shown in Fig. 151 are supposed to illustrate successive stages in the development of an average star. Type I corresponds to the period in which its temperature is near the maximum; Type II belongs to a later stage in which the temperature has commenced to fall; and Type III to the period immediately preceding extinction.

While human life, or even the duration of the human race, is too short to permit a single star to be followed through all the stages of its career, an adequate picture of that development might be obtained by examining many stars, each at a different stage of progress, and, following this idea, numerous subdivisions of the types of stellar spectra shown in Fig. 151 have been proposed in order to represent with more detail the process of stellar growth and decay; but for the most part these subdivisions and their interpretation are accepted by astronomers with much reserve.

It is significant that there are comparatively few stars with spectra of Type III, for this is what we should expect to find if the development of a star through the last stages of its visible career occupied but a small fraction of its total life. From the same point of view the great number of stars with spectra of the first type would point to a long duration of this stage of life. The period in which the sun belongs, represented by Type II, probably has a duration intermediate between the others. Since most of the variable stars, save those of the Algol class, have spectra of the third type, we conclude that variability, with its associated ruddy color and great atmospheric absorption of light, is a sign of old age and approaching extinction. The Algol or eclipse variables, on the other hand, having spectra of the first type, are comparatively young stars, and, as we shall see a little later, the shortness of their light periods in some measure confirms this conclusion drawn from their spectra.

We have noted in § 196 that the sun's near neighbors are prevailingly stars with spectra of the second type, while the Milky Way is mainly composed of first-type stars, and from this we may now conclude that in our particular part of the entire celestial space the stars are, as a rule, somewhat further developed than is the case elsewhere.

244. ***Double stars.***— The double stars present special problems of development growing out of the effects of tidal friction, which must operate in them much as it does between earth and moon, tending steadily to increase the distance between the components of such a star. So, too, in such a system as is shown in Fig. 132, gravity must tend to make each component of the double star shrink to smaller dimensions, and this shrinkage must result in faster rotation and increased tidal friction, which in turn must push the components apart, so that in view of the small density and close proximity of those particular stars we may fairly regard a star like β Lyræ as in the early stages of its career and destined with increasing age to lose its variability of light, since the eclipses which now take place must cease with increasing distance between the components unless the orbit is turned exactly edgewise toward the earth. Close proximity and the resulting shortness of periodic time in a double star seem, therefore, to be evidence of its youth, and since this shortness of periodic time is characteristic of both Algol variables and spectroscopic binaries as a class, we may set them down as being, upon the whole, stars in the early stages of their career. On the other hand, it is generally true that the larger the orbit, and the greater the periodic time in the orbit, the farther is the star advanced in its development.

In his theory of tidal friction, Darwin has pointed out that whenever the periodic time in the orbit is more than twice as long as the time required for rotation about the axis, the effect of the tides is to increase the eccentricity of the orbit, and, following this indication, See has urged that with increasing distance between the components of a double star their orbits about the common center of gravity must grow more and more eccentric, so that we have in the shape of such orbits a new index of stellar development; the more eccentric the orbit, the farther advanced are the stars. It is important to note in this connection that among the double stars whose orbits have been computed there seems to run a general rule— the larger the orbit the greater is its eccentricity— a relation which must hold true if tidal friction operates as above supposed, and which, being found to hold true, confirms in some degree the criteria of stellar age which are furnished by the theory of tidal friction.

245. ***Nebulæ.***— The nebular hypothesis of Laplace has inclined astronomers to look upon nebulæ in general as material destined to be worked up into stars, but which is now in a very crude and undeveloped stage. Their great bulk and small density seem also to indicate that gravitation has not yet produced in them results at all comparable with what we see in sun and stars. But even among nebulæ there are to be found very different stages of development. The irregular nebula, shapeless and void like that of Orion; the spiral, ring, and planetary nebulæ and the star cluster, clearly differ in amount of progress toward their final goal. But it is by no means sure that these several types are different stages in one line of development; for example, the primitive nebula which grows into a spiral may never become a ring or planetary nebula, and *vice versa.* So too there is no reason to suppose that a star cluster will ever break up into isolated stars such as those whose relation to each other is shown in Fig. 122.

246. ***Classification.***— Considering the heavenly bodies with respect to their stage of development, and arranging them in due order, we should probably find lowest down in the scale of progress the irregular nebulæ of chaotic appearance such as that represented in Fig. 146. Above these in point of development stand the spiral, ring, and planetary nebulæ, although the exact sequence in which they should be arranged remains a matter of doubt. Still higher up in the scale are star clusters whose individual members, as well as isolated stars, are to be classified by means of their spectra, as shown in Fig. 151, where the order of development of each star is probably from Type I, through II, into III and beyond, to extinction of its light and the cutting off of most of its radiant energy. Jupiter and Saturn are to be regarded as stars which have recently entered this dark stage. The earth is further developed than these, but it is not so far along as are Mars and Mercury; while the moon is to be looked upon as the most advanced heavenly body accessible to our research, having reached a state of decrepitude which may

almost be called death— a stage typical of that toward which all the others are moving.

Meteors and comets are to be regarded as fragments of celestial matter, chips, too small to achieve by themselves much progress along the normal lines of development, but destined sooner or later, by collision with some larger body, to share thenceforth in its fortunes.

247. ***Stability of the universe.*** — It was considered a great achievement in the mathematical astronomy of a century ago when Laplace showed that the mutual attractions of sun and planets might indeed produce endless perturbations in the motions and positions of these bodies, but could never bring about collisions among them or greatly alter their existing orbits. But in the proof of this great theorem two influences were neglected, either of which is fatal to its validity. One of these— tidal friction— as we have already seen, tends to wreck the systems of satellites, and the same effect must be produced upon the planets by any other influence which tends to impede their orbital motion. It is the inertia of the planet in its forward movement that balances the sun's attraction, and any diminution of the planet's velocity will give this attraction the upper hand and must ultimately precipitate the planet into the sun. The meteoric matter with which the earth comes ceaselessly into collision must have just this influence, although its effects are very small, and something of the same kind may come from the medium which transmits radiant energy through the interstellar spaces.

It seems incredible that the luminiferous ether, which is supposed to pervade all space, should present absolutely no resistance to the motion of stars and planets rushing through it with velocities which in many cases exceed 50,000 miles per hour. If there is a resistance to this motion, however small, we may extend to the whole visible universe the words of Thomson and Tait, who say in their great Treatise on Natural Philosophy, "We have no data in the present state of science for estimating the relative importance of tidal friction and of the resistance of the resisting medium through which the earth and moon move; but, whatever it may be, there can be but one ultimate result for such a system as that of the sun and planets, if continuing long enough under existing laws and not disturbed by meeting with other moving masses in space. That result is the falling together of all into one mass, which, although rotating for a time, must in the end come to rest relatively to the surrounding medium."

Compare with this the words of a great poet who in The Tempest puts into the mouth of Prospero the lines:

"The cloud-capp'd towers, the gorgeous palaces,The solemn temples, the great globe itself,Yea, all which it inherit, shall dissolve;And, like this insubstantial pageant faded,Leave not a rack behind."

248. ***The future.*** — In spite of statements like these, it lies beyond the scope of scientific research to affirm that the visible order of things will ever

come to naught, and the outcome of present tendencies, as sketched above, may be profoundly modified in ages to come, by influences of which we are now ignorant. We have already noted that the farther our speculation extends into either past or future, the more insecure are its conclusions, and the remoter consequences of present laws are to be accepted with a corresponding reserve. But the one great fact which stands out clear in this connection is that of *change*. The old concept of a universe created in finished form and destined so to abide until its final dissolution, has passed away from scientific thought and is replaced by the idea of slow development. A universe which is ever becoming something else and is never finished, as shadowed forth by Goethe in the lines:

"Thus work I at the roaring loom of Time,And weave for Deity a living robe sublime."

FOOTNOTES

[A]The circle and straight line are considered to be special cases of these curves, which, taken collectively, are called the conic sections.

[B]Aristophanes, The Clouds, Whewell's translation.

[C]Schiaparelli, Osservazioni sulle Stelle Doppie.

APPENDIX

The Greek Alphabet

The Greek letters are so much used by astronomers in connection with the names of the stars, and for other purposes, that the Greek alphabet is printed below— not necessarily to be learned, but for convenient reference:

Greek.		*Name.*	*English.*
Α	α	Alpha	a
Β	β	Beta	b
Γ	γ	Gamma	g
Δ	δ	Delta	d
Ε	ε or ϵ	Epsilon	ĕ
Ζ	ζ	Zeta	z
Η	η	Eta	ē
Θ	ϑ or θ	Theta	th
Ι	ι	Iota	i
Κ	κ	Kappa	k
Λ	λ	Lambda	l
Μ	μ	Mu	m
Ν	ν	Nu	n
Ξ	ξ	Xi	x
Ο	ο	Omicron	ŏ
Π	π	Pi	p
Ρ	ρ	Rho	r
Σ	σ or ς	Sigma	s
Τ	τ	Tau	t
Υ	υ	Upsilon	u

Φ	φ	Phi	ph
Χ	χ	Chi	ch
Ψ	ψ	Psi	ps
Ω	ω	Omega	ō

Popular Literature of Astronomy

The following brief bibliography, while making no pretense at completeness, may serve as a useful guide to supplementary reading:

General Treatises

YOUNG. *General Astronomy.* An admirable general survey of the entire field.

NEWCOMB. *Popular Astronomy.* The second edition of a German translation of this work by Engelmann and Vogel is especially valuable.

BALL. *Story of the Heavens.* Somewhat easier reading than either of the preceding.

CHAMBERS. *Descriptive Astronomy.* An elaborate but elementary work in three volumes.

LANGLEY. *The New Astronomy.* Treats mainly of the physical condition of the celestial bodies.

PROCTOR and RANYARD. *Old and New Astronomy.*

Special Treatises

PROCTOR. *The Moon.* A general treatment of the subject.

NASMYTH and CARPENTER. *The Moon.* An admirably illustrated but expensive work dealing mainly with the topography and physical conditions of the moon. There is a cheaper and very good edition in German.

YOUNG. *The Sun.* International Scientific Series. The most recent and authoritative treatise on this subject.

PROCTOR. *Other Worlds than Ours.* An account of planets, comets, etc.

NEWTON. *Meteor.* Encyclopædia Britannica.

AIRY. *Gravitation.* A non-mathematical exposition of the laws of planetary motion.

STOKES. *On Light as a Means of Investigation.* Burnett Lectures. II. The basis of spectrum analysis.

SCHELLEN. *Spectrum Analysis.*

THOMSON (Sir W., Lord KELVIN), *Popular Lectures, etc.* Lectures on the Tides, The Sun's Heat, etc.

BALL. *Time and Tide.* An exposition of the researches of G. H. Darwin upon tidal friction.

GORE. *The Visible Universe.* Deals with a class of problems inadequately treated in most popular astronomies.

DARWIN. *The Tides.* An admirable elementary exposition.

CLERKE. *The System of the Stars.* Stellar astronomy.

NEWCOMB. Chapters on the Stars, in *Popular Science Monthly* for 1900.

CLERKE. *History of Astronomy during the Nineteenth Century.* An admirable work.

WOLF. *Geschichte der Astronomie.* München, 1877. An excellent German work.

A List of Stars for Time Observations

See § 20.

NAME.	*Magnitude.*	*Right Ascension.*		*Declination.*
		h.	m.	°
β Ceti	2	0	38.6	-18.5
η Ceti	3	1	3.6	-10.7
α Ceti	3	2	57.1	+3.7
γ Eridani	3	3	53.4	-13.8
Aldebaran	1	4	30.2	+16.3
Rigel	0	5	9.7	-8.3
κ Orionis	2	5	43.0	-9.7
β Canis Majoris	2	6	18.3	-17.9
Sirius	-1	6	40.7	-16.6
Procyon	0	7	34.1	+5.5
α Hydræ	2	9	22.7	-8.2

NAME.	*Magnitude.*	*Right Ascension.*		*Declination.*
Regulus	1	10	3.0	+12.5
ν Hydræ	3	10	44.7	-15.7
ϵ Corvi	3	12	5.0	-22.1
γ Corvi	3	12	10.7	-17.0
Spica	1	13	19.9	-10.6
ζ Virginis	3	13	29.6	-0.1
α Libræ	3	14	45.3	-15.6
β Libræ	3	15	11.6	-9.0
Antares	1	16	23.3	-26.2
α Ophiuchi	2	17	30.3	+12.6
ϵ Sagittarii	2	18	17.5	-34.4
δ Aquilæ	3	19	20.5	+2.9
Altair	1	19	45.9	+8.6
β Aquarii	3	21	26.3	-6.0
α Aquarii	3	22	0.6	-0.8
Fomalhaut	1	22	52.1	-30.2

THE END

Detailed Historical Context

INTRODUCTION

The nineteenth century was a fascinating and vital formative period in Western literature since it provided the fundamental backdrop for the formation and emergence of contemporary literary traditions and styles as we know them today.

The Victorian Period, named after Queen Victoria's reign from 1837 to 1901, was characterised by significant cultural and creative triumphs, social and technological developments, and significant political and economic transformation. It was a time of development and expansion for Britain, as it became the world's largest empire; it was also a time of significant social and cultural transformation in America. Rapid industrialisation and urbanisation resulted in a lively literary environment with a diverse spectrum of genres and styles. Popular literary genres at the time included sentimental novels, gothic novels, and regionalist writing. Additional research, for example, shows that the Romantic, Symbolist, and Realist movements, as well as a variety of social and economic circumstances that dominated the twentieth century, all had their origins and predecessors in the nineteenth century.

MOVEMENTS AND LITERATURE

ROMANTICISM

Romanticism, with its stress on sensation and the irrational, emerged in the nineteenth century as a significant literary and cultural movement. The 18th century, on the other hand, was regarded to be the age of intelligence, reasoning, and the mind. Romanticism, which emerged from the late-nineteenth-century German Sturm und Drang ("Stress and Storm") movement and whose notable members included Goethe and Friedrich Schiller, was marked by a focus on the individual, subjective, mystical, emotional, and inner life.

Writers and poets such as William Wordsworth, Samuel Taylor Coleridge, and John Keats in England, and Johann Wolfgang von Goethe and Friedrich Schiller in Germany, sought to capture the sublime in nature and the depth of human emotion in their works.

The Romantic movement was also marked by a fascination with the past, the mystical, and the exotic. This was evident in the rise of Gothic literature, with novels such as Mary Shelley's 'Frankenstein' (1818) and the poems of Edgar Allan Poe. Romanticism was not just a literary movement; it also had profound impacts on art and music, inspiring artists like J.M.W. Turner and composers like Ludwig van Beethoven. Ultimately, Romanticism represented a fundamental shift in cultural attitudes, offering a new perspective on the nature of creativity, the purpose of art, and the role of the artist in society.

- *Rousseau*

Jean-Jacques Rousseau was a towering intellectual figure whose ideas shaped the 19th century, despite his death in 1778. His writings profoundly influenced both the Age of Enlightenment and the Romantic movements, creating a bridge between these two key periods. Rousseau challenged the primacy of reason advocated by his Enlightenment contemporaries, arguing that feelings and emotions were also essential in understanding the human experience. His novel, "Julie, or the New Heloise" (1761), is considered a precursor to Romanticism, emphasizing passion and sentiment.

- *Early Romantic poets*

The late 18th and early 19th century Romantic English poets William Wordsworth and Samuel Taylor Coleridge, who released their collection of poems Lyrical Ballads in 1798, are considered the forefathers of this style. As seen by the works of Pushkin in Russia, Ugo Foscolo and Giacomo Leopardi in Italy, José de Espronceda in Spain, and Giacomo Leopardi, the Romantic poetry movement was popular and flourished throughout Europe and beyond.

- *American Romanticism*

American Romanticism, a movement that spanned the mid-19th century, was a reaction against the rationalism of the Age of Enlightenment and a manifestation of the ethos of individualism that was central to the American frontier spirit. It encapsulated a broad range of human experience and played out differently across various genres, exploring themes like the supernatural, the power of nature, and the potential of the individual.

James Fenimore Cooper's historical adventure novels, such as "The Last of the Mohicans" (1826), created a uniquely American kind of Romantic hero - the rugged, self-reliant frontiersman. Edgar Allan Poe took a darker route, delving into the eerie and supernatural in tales like "The Fall of the House of Usher" (1839) and "The Raven" (1845). These works were representative of the Gothic element within Romanticism, exploring the darker recesses of the human psyche.

Walt Whitman, with his groundbreaking collection "Leaves of Grass" (1855), embodied another aspect of American Romanticism. His free verse celebrated the individual, democratic values, and the spiritual significance of everyday life.

Finally, the Transcendentalist movement, led by Ralph Waldo Emerson and Henry David Thoreau, elevated the individual conscience above societal norms. Emerson's essay "Self-Reliance" (1841) became a key text, while Thoreau's "Walden" (1854) documented his experiment in simple living and immersion in nature.

Together, these authors and works shaped the American Romantic movement, offering new perspectives on the human experience and inspiring readers to break free from societal constraints and explore their own individual paths.

- *Second Generation Romantic poets*

To discover the "truth" of things, the Romantics went to people's emotions, which were grounded in and exemplified by interaction with nature and the primordial self, rather than logical inquiry. Second-generation Romantic writers John Keats, Lord Byron, and Percy Bysshe Shelley's writings are good examples of these points of view.

POST ROMANTICISM

- *Parnassianism*

The works of French poets Théophile Gautier and Charles Baudelaire are examples of Parnassianism, which can be considered as an extension of early Romantic viewpoints with its emphasis on aesthetics and the concept of art for the sake of art. Schopenhauer's philosophical ideas had an impact as well. Devotees attempted to address their foreign and old subjects of fascination in a more controlled, formal manner, retreating from the excess passion and sentimentality of the Romantic movement.

- *Impressionism and Symbolism*

Claude Monet and other Paris-based painters contributed to the development of impressionism, which first arose in painting and then in music in France near the end of the nineteenth century. Impressionism was a painting style that attempted to reflect the visual world as accurately as possible by employing the shifting qualities of light and colour as seen via human perception and experience.

Symbolism is characterised as a departure from naturalism and realism in favour of a harsher, more truthful portrayal of the world, with a concentration on the ordinary rather than the extraordinary. Symbolist poets, such as Gustave Kahn and Ezra Pound, employed imagery to "evoke" rather than portray or describe.

THE GOTHIC NOVEL

The Gothic Novel, a vibrant subgenre of Romantic fiction, emerged in Europe towards the end of the 18th century. Pioneers in this field include Horace Walpole with his ground-breaking novel "The Castle of Otranto" (1765), and Ann Radcliffe, whose work "The Mysteries of Udolpho" elevated the genre. The term 'Gothic' is derived from Gothic architecture, a common setting in these novels, characterized by crumbling castles, haunted monasteries, and dark forests, which lent an eerie atmosphere to the narratives.

Distinct from typical supernatural tales, Gothic novels often dealt with themes of ancestral curses and past sins haunting the present, exploring the darker recesses of the human psyche and the effects of terror and horror on it. This was further explored in the 19th-century through seminal works such as Mary Shelley's "Frankenstein" (1818), Sir Walter Scott's "Bride of Lammermoor" (1819), E.T.A. Hoffmann's "The Devil's Elixirs" (1815), Emily Bronte's "Wuthering Heights" (1847), Robert Louis Stevenson's "The Strange Case of Dr. Jekyll and Mr. Hyde" (1886), and Bram Stoker's "Dracula" (1897).

Even beyond these iconic pieces, the influence of the Gothic novel can be seen in many well-regarded Victorian works. Charles Dickens' "Bleak House" (1852-1853) and "Great Expectations" (1861), for example, both incorporate elements of the Gothic tradition, reflecting its broad impact on the literature of the time. This genre, with its exploration of the sublime, the uncanny, and the spectral, significantly contributed to the richness and depth of 19th-century literature.

POPULAR PHILOSOPHY

- *German Idealism*

German Idealism, a significant philosophical movement of the late 18th and early 19th centuries, was pioneered by figures such as Johann Gottlieb Fichte. Building upon the metaphysical insights of Immanuel Kant, Fichte proposed a dynamic conception of the self as a constantly evolving entity. Georg Wilhelm Friedrich Hegel further extended this idea by emphasizing the importance of historical and dialectical thinking in understanding the self.

In contrast, Arthur Schopenhauer diverged from Hegel's path and argued for a return to Kant's transcendental philosophy.

- *Marxism*

The philosophical and political ideology known as Marxism was born out of the intellectual partnership of Karl Marx and Friedrich Engels. Their seminal work, "The Communist Manifesto" (1848), presented a critique of capitalism, asserting its inherent instability and predicting its eventual replacement by socialist and, subsequently, communist systems. This work laid the foundation for the later international communist movement.

- *Positivism*

The philosophical position of Positivism was proposed by August Comte, advocating the belief that genuine knowledge is inherently empirical and verifiable. Comte argued that such knowledge derives from observable phenomena and subsequent logical and mathematical reasoning, excluding innate knowledge or metaphysical speculation

- *Social Darwinism*

The concept of Social Darwinism sought to apply the biological principles of natural selection and survival of the fittest, as outlined in Charles Darwin's "On the Origin of Species", to societal and political contexts. Proponents included Francis Galton, who maintained that cognitive abilities were as heritable as physical characteristics, and advocated for societal intervention in reproductive practices to prevent the over-breeding of "less fit" individuals. Similarly, Herbert Spencer, in his work "The Social Organism" (1860), likened society to a living organism, evolving and adapting according to Darwinian principles.

SOCIAL, ECONOMIC AND POLITICAL IMPACTS

- *The Industrial Revolution*

The Industrial Revolution, which took place between the late 18th and a time between 1820 and 1840, was a time of great social, political, and economic uprisings and change that involved the challenging transition from largely manual production methods to mechanical manufacturing methods, particularly in the fields of textiles, steam power, iron making, and the invention of machine tools. Agriculture had previously been the foundation of the European economy, and it was also a time when basic political, scientific, and religious ideas were unravelled to their core.

As a result of this mechanisation, a considerable number of people were transported from rural villages to metropolitan regions, resulting in a significant increase in population and the establishment of new, larger cities. The advancement of new technology resulted in the establishment of factories, a dehumanising and horrifying method of labour, particularly child labour, and a capitalist way of life. Because cities were unable to accommodate the rapidly rising population, there were overcrowded slums and terribly deplorable living conditions, as described in books such as Friedrich Engels' The Condition of the Working Class in England, published in 1844.

Elizabeth Barrett Browning's The Cry of the Children, Thomas Hardy's Tess of the D'Urbervilles, and works by author and philosopher Thomas Carlyle warned of the threat to society posed by these inhumane conditions and the profit-focused, materialistic ideals of what Dickens referred to as the "mechanical age" in his novels Hard Times and Oliver Twist.

- *Slavery and the Abolionist movement*

The 19th century in the United States was a time of great political upheaval and moral conflict. At the heart of these struggles was the question of slavery - the practice of owning human beings as property and forcing them to labor for the benefit of their owners. Slavery was deeply entrenched in the southern states, where it was seen as essential to the region's economy and way of life. But in the north, a growing abolitionist movement called for the immediate and unconditional end of slavery, seeing it as a fundamental violation of human rights and a stain on the nation's conscience.

These debates over slavery were not just academic or theoretical - they were deeply intertwined with the politics and culture of the time. The question of whether or not to allow slavery in new territories was a key issue in the lead-up to the Civil War, which ultimately erupted in 1861 and tore the nation apart. But even before the war, tensions over slavery were high, and political leaders grappled with how to address this thorny and divisive issue.

While the United States grappled with the practice and morality of slavery within its own borders, across the Atlantic, the United Kingdom was undergoing its own transformation in the 19th century regarding slavery. In 1807, the UK took a decisive step with the passage of the Slave Trade Act, which outlawed the transatlantic slave trade. This was followed by the Slavery Abolition Act of 1833, effectively ending slavery throughout the British Empire, except for areas under the administration of the East India

Company, and in the territories of Ceylon (now Sri Lanka) and Saint Helena. This Act marked a critical turning point in the global fight against slavery.

This momentous development was not without its influences. Several influential works published during this period galvanized public opinion and shaped the discourse on slavery and abolition. Thomas Clarkson's 'An Essay on the Slavery and Commerce of the Human Species' (1786) offered a thorough critique of slavery, leading to its expanded edition in 1808. Another influential work, 'The History of Mary Prince' (1831), was the first account of a black woman's life published in the UK, detailing her experiences as an enslaved person in Bermuda, which sparked public interest and became a tool in the hands of abolitionists.

In tandem, anti-slavery sentiment was reflected in the literary world as well. For instance, Elizabeth Barrett Browning's influential poem, 'The Runaway Slave at Pilgrim's Point' (1847), powerfully condemned the institution of slavery.

- *The Rise of Nationalism and Imperialism*

The 19th century marked a pivotal period in global history, as it saw the rise of two influential ideologies: Nationalism and Imperialism. Both had profound implications for the world, reshaping political, economic, and social landscapes.

Nationalism emerged as a potent political force, rooted in the belief that individuals sharing a common language, culture, or ancestry constituted a nation. This ideology played a critical role in the unification of fragmented regions into cohesive nation-states. The unification of Italy in 1861 and Germany in 1871 stand as two of the most significant examples of nationalism's impact. Both unifications were driven by charismatic leaders— Camilo di Cavour in Italy and Otto von Bismarck in Germany— and the shared desire of the people to form a unified national identity. The emergence of nationalism also led to a rise in independence movements in various parts of the world, leading to the downfall of old empires and the birth of new nations.

Imperialism, on the other hand, was driven by the ambitions of the powerful Western nations to expand their influence and control over other parts of the globe. Rooted in a belief in cultural and racial superiority, as well as economic motivations, imperialism led to the colonization of large parts of Africa, Asia, and the Pacific. Key events during this period include the scramble for Africa (1881-1914), where European powers divided the continent among themselves, and the Opium Wars (1839-1860), which marked the beginning of Western imperial control over China.

The expansion of the British Empire, which, at its height, was the largest empire in history, is another prominent example of 19th-century imperialism. This period also witnessed the rise of the United States as an imperial power, with its acquisition of territories in the Caribbean and Pacific, notably following the Spanish-American War in 1898.

Both nationalism and imperialism had profound and lasting impacts on global politics, economics, and societies, the effects of which continue to be felt into the present day. They shaped national identities, redrawed the world map, and sowed the seeds for many of the conflicts and power dynamics of the 20th century.

- *Science and influential Non-Fiction works*

Throughout the nineteenth century, Victorians' drive to understand and categorise the natural world played an important part in the development of scientific theory and understanding. Charles Darwin's works, such as the well-known On the Origin of Species (1859), would have a dramatic and far-reaching impact due to their innovative idea of evolution, which contradicted many of the time's established notions and religious beliefs.

The French Revolution: A History, published in 1837, and On Heroes, Hero-Worship, and the Heroic in History, published in 1841, are two other important non-fiction works from the period that influenced political thinking in the mid-nineteenth century.

- *Scientific Perspectives (20th Century)*

The advancements in science, like Albert Einstein's theory of relativity, brought a different viewpoint to look at the world. This also had a profound effect on literature, with many authors incorporating these concepts into their works. Similarly, the ideas of Sigmund Freud and Carl Jung about the human mind influenced writers to explore characters' inner lives and subconscious in depth.

KEY HISTORICAL EVENTS

- *The Acts of Union and Treaty of Amiens*

Following the French Revolution and the Irish Rebellion in the late 18th century, which brought unpredictability and instability, the Acts of Union were passed in 1800, combining Britain and Ireland to form the United Kingdom.

The Treaty of Amiens brought the Second Coalition French Revolutionary War to a close, as well as the disputes between France and the United Kingdom. The effect, however, was fleeting, since the Napoleonic Wars began in just three years.

- *US expansion*

With their newly gained independence, the United States chose to purchase the Louisiana Purchase in 1803 in order to double their size and expand their control over the Mississippi River. Native Americans occupy a substantial portion of the region, which was purchased for $15 million from the French First Republic.

- *Napoleonic Wars*

Napoleon decimated the Russian and Austrian troops in 1805, but his plans to invade England were thwarted when Admiral Nelson soundly defeated the French and Spanish armies at the Battle of Trafalgar, cementing the nation's dominance over the oceans.

During the Russian invasion, the French army suffered tremendous losses– up to 380,000 soldiers died– and Napoleon's previous image as an unbeatable general was shattered. The French king abdicated and was exiled to Elba following his loss in the War of the Sixth Coalition in 1814.

- *British and Russian empire expansion*

Following France's loss, Britain and Russia rose to prominence as the world's two largest powers, with Russia expanding its sphere of influence to include Central Asia and the Caucasus and Britain increasing its foreign possessions to include Canada, Australia, South Africa, and Africa. The British East India Company was dissolved as a result of the Indian Rebellion of 1857, which was a widespread insurrection against colonial rule. Later, the British Crown assumed direct administration and founded the British Raj.

- *Opium wars*

By the mid-nineteenth century, China experienced severe opium problems as a result of the opening of trade with the West and the illicit trafficking in the drug coordinated by British entrepreneurs seeking to earn money at the trading ports. On the basis of free trade principles, Britain resisted the emperor's attempt to outlaw its sale, resulting in the First Opium War and the Treaty of Nanking in 1842, which permitted the drug trade to continue while handing over control of Hong Kong to the British.

The Taiping Rebellion of 1856 set the stage for the second Opium War, in which France and Britain collaborated. The 1860 Peking Convention, which legalised the opium trade and forced the surrender of additional provinces, resulted in the early nineteenth-century demise of the Qing dynasty.

- *The 1848 Revolutions*

The 1848 Revolutions, also known as the Springtime of the Nations, were a series of political upheavals that occurred in Europe and the rest of the world in 1848. The main purpose of these uprisings was to abolish previous monarchical authority and establish free nation governments.

Nationalists in Italy organised revolutions in Sicily and the Italian peninsula republics in order to construct a liberal government and break free from Austrian domination. The February Revolution in France occurred in Paris following the crackdown on the campagne des banquets, a violent insurgency against the monarchy that resulted in King Louis Philippe's overthrow. Germany, Denmark, Hungary, Galicia, Sweden, and Switzerland were among the other countries that revolted against the Habsburg Monarchy.

- *Abolitionism and the end of slavery*

The Atlantic slave trade was outlawed in the United States in 1808, and slavery was outlawed throughout the British Empire by the Slavery Abolition Act of 1833. Abolitionism triumphed in the nineteenth century. Abolitionism persisted in the United States until the Civil War ended in 1865, when the Thirteenth Amendment to the Constitution was ratified, officially abolishing slavery in the country.

- *Women's Suffrage movement*

"The call for women's suffrage, the right for women to vote, gained momentum in the 19th century. The cause was propelled by landmark events such as the Seneca Falls Convention in 1848, often considered the birthplace of the American women's rights movement, and the first National Women's Rights Convention in 1850. Key feminist literary works, such as Margaret Fuller's 'Woman in the Nineteenth Century' (1845), which advocated for women's independence and equality, and Sarah Grimké's 'The Equality of the Sexes and the Condition of Women' (1838), helped shape the discourse surrounding women's rights.

In spite of initial failures to secure voting rights during the 1870s, suffragists pressed forward with relentless determination, advocating state-by-state for a constitutional amendment to enfranchise women. The U.S. state of Wyoming led the way in 1869, becoming the first state to grant women the right to vote in all elections.

Across the Atlantic, the UK suffragette movement found a dynamic leader in Emmeline Pankhurst. Pankhurst, who became involved in the suffrage movement in the 1880s, was instrumental in shifting the strategy of the movement towards direct action and civil disobedience, a strategy that would ultimately lead to women achieving the right to vote in 1918 for women over 30 and in 1928 for all women over 21, leveling the voting age with men."

INTRODUCTION- 20th Century

The twentieth century saw significant upheaval in both society and literature. The two World Wars, the Great Depression, which affected the entire world economy, the fall of the British Empire and the subsequent changes in world politics, as well as ongoing technological and scientific advances, all contributed to an almost unprecedented period of cataclysmic change in human history.

The century was defined by two significant literary periods: the Modernist Period (1910–1945) and the Contemporary or Postmodern Period (after 1945). These periods in Britain and Europe were distinguished by creative and experimental kinds of writing that questioned standard literary rules. In America, writers such as Robert Frost and Flannery O'Connor experimented with new forms and styles of writing during the Modernist Period, while The Lost Generation (also known as The Jazz Age, 1914-1929) saw famous writers such as Hemingway, Stein, Fitzgerald, and Faulkner experiment with new forms of storytelling.

The Modernist Period saw writers such as W. B. Yeats, Seamus Heaney, Dylan Thomas, W. H. Auden, Virginia Woolf, and Wilfred Owen experiment with new and unconventional genres of poetry, fiction, and theatre. These writers experimented with fragmented and stream-of-consciousness narratives in order to break free from standard forms and norms.

Following World War I, a new unsettling and challenging type of poetry evolved, Imagism and Modernism emerged in the interwar period, and the post-World War II period saw the emergence of many reactions to Modernism, such as the Postmodernism movement. Leaving the Victorian era behind, the Georgian and Aesthetic movement poets emerged around the turn of the century.

Following World War II, the Postmodern Period saw a further break from established forms and conventions. T. S. Eliot, Morrison, Shaw, Beckett, Stoppard, Fowles, Calvino, Ginsberg, Pynchon, and other postmodern writers experimented with metafiction, fragmented poetry, and other kinds of writing that challenged the boundaries between fiction and reality.

MOVEMENTS AND LITERATURE

THE EDWARDIANS

By the turn of the century, marked by Queen Victoria's death in 1901 and the accession of Edward VII, there was a sense of dissatisfaction with and a move away from the previous inflexible Victorian certainties and conservatism. Among the influential philosophical, scientific, and political thinkers of the time were Albert Einstein, Charles Darwin, Sigmund Freud, Friedrich Nietzsche, and Karl Marx, whose works had a profound and revolutionary impact on Western culture, beliefs, and humanity's understanding of itself and its origins.

- *H.G. Wells, modernism and a new era*

As a new century and an uncertain future arrived, people were both excited for the future and concerned about what change might bring. In his early Utopian works, such as Anticipations of the Reaction of Mechanical and Scientific Progress upon Human Life and Thought (1901), Mankind in the Making (1903), and Utopia (1905), English writer H.G. Wells explored and anticipated the effects and transformations that science and technological advancement could bring.

- *Realism and a new role for the arts*

With its motto of "Art for Art's Sake," which was championed by Decadent authors and painters near the end of the nineteenth century, pure Aestheticism was something that many writers, including Dante Gabriel Rossetti, Oscar Wilde, and Algernon Charles Swinburne, purposefully moved away from throughout the Edwardian era. Writers, particularly those of dramatic works, have started to use their works as a forum for discussion and study of the day's major moral and social concerns, touching on a wide range of topics such as politics, the role of women, marriage as a social institution, and the ethics of war.

Playwright George Bernard Shaw's comic pieces, Man and Superman (1903) and Major Barbara (1907), John Galsworthy's examination of industry morals and politics in Strife (1909), and Harley Granville Barker's critique of class hypocrisy in The Voysey Inheritance (1905), were all heavily influenced by the realism and naturalistic techniques of 19th-century dramatists and writers like Ibsen, Balzac, and Dickens.

- *Transition and traditionalism*

Thomas Hardy is an example of a writer who, like Rudyard Kipling and G.K. Chesterton, achieved literary success in the previous century. He might be seen as a transitional character who spanned the gap between the Victorian and Edwardian eras by attempting to retain conventional literary styles and techniques. Similar attempts to resurrect traditional techniques and topics can be found in A.E. Housman's pastoral poetry. The Georgian poets, who included Rupert Brooke, Robert Graves, and Edmund Blunden in the early twentieth century, maintained a more conventional, moderate poetic style by continuing to use romantic and emotional methods.

- *Henry James and Joseph Conrad*

Another major transitional character in literature who helped bridge the gap between realism and modernism is Henry James, a British novelist of American heritage whose writings are famous for focusing on the collision between the Old World and the New, as well as between Americans and Europeans. His final three writings, The Wings of the Dove, The Ambassadors, and The Golden Bowl, published between 1902 and 1904, were filled with dread and despair about a new world order in which ancient social institutions and moral certitudes were vanishing. Joseph Conrad analyses the cost of human arrogance and failures in innovative works such as Heart of Darkness (1902), Nostromo (1907), and Under Western Eyes (1911), expressing a similar dismay with the world.

MODERNISM

The early 20th century was indeed a time of great change and innovation in the literary world. This period saw a distinct shift away from the 19th-century Romantic and Victorian traditions.

The most significant literary movement of the early 20th century was Modernism. This was characterized by a self-conscious break with traditional ways of writing, in both poetry and prose. Literary modernism, or modernist literature, originated from the late 19th and early 20th centuries, broadly speaking, with the advent of World War I. This period saw an increase in the use of stream-of-consciousness narrative, a technique that seeks to depict the multitudinous thoughts and feelings which pass through the mind. Another striking characteristic of Modernism is the use of fragmentation. The world was perceived as disjointed and chaotic, hence the writing showed that through its structure.

- *Imagist movement*

The work of the imagists, a group of English and American poets united under the banner of ideals advocated by Ezra Pound in his poetry anthologies Ripostes (1912) and Des Imagistes (1914), is an excellent example of this new modernist mindset. T. E. Hulme, Richard Adlington, Hilda Doolittle, and Amy Lowell were all important members. In their experimental poetry, they used free verse and other unconventional forms.

- *The modernist novel*

The tone and subject matter of literary and lyrical works created after World War I began in 1914 reflect the war's devastation. D.H. Lawrence's modernist worldview can be observed in his extraordinary use of a stream-of-consciousness style, as seen in The Rainbow (1915) and Women in Love (1920), as well as his assessment of the disastrous ramifications of industrialisation. Ulysses, published in 1922 by James Joyce, also adopted the stream-of-consciousness style and effectively defined the modernist literary movement as a whole.

- *World War I Literature*

Among the poets whose works were inspired by World War I were Rupert Brooke, whose patriotic poem 1914 successfully evoked the idealistic nature of the conflict's early months, Wilfred Owen, whose poems powerfully depict both the comradeship and the moral dilemmas of war, and Siegfried Sassoon, who captured the mounting feelings of rage and wastefulness as the war raged on. Many of these works, however, were not well-known or recognised until the 1930s.

- *T.S. Eliot and The Waste Land*

T.S. Eliot is a modernist poet recognised for his use of experimental language and style. Through the use of a disconnected style and dissonant allusion to a wide range of literary and cultural works, Eliot masterfully captured the disillusionment and hostility experienced from the perspective of the post-war period in two of his most renowned works, Prufrock and Other Observations (1917) and The Waste Land (1922). He recognised spiritual emptiness as the source of the psychological sickness afflicting civilisation today, yet he nonetheless offered alternatives for healing and rebirth through mythology and symbolism.

- *The Harlem Renaissance*

The Harlem Renaissance was an important cultural movement that saw the growth of black writers like Baldwin and Ellison. W.E.B. Du Bois and other black writers and activists published slave narratives and books that provided profound insight into the black experience.

They aimed to undermine standard narrative assumptions and question accepted literary interpretations.

- *Scientific Perspectives*

The advancements in science, like Albert Einstein's theory of relativity, brought a different viewpoint to look at the world. This also had a profound effect on literature, with many authors incorporating these concepts into their works. Similarly, the ideas of Sigmund Freud and Carl Jung about the human mind influenced writers to explore characters' inner lives and subconscious in depth.

- *Literature in the 1920s and 1930s*

The works of modernist writers like as Eugene O'Neill, whose experimental techniques and use of naturalism and expressionist practises in his theatrical performances were enormously influential at the time, aided American theatre in reaching its zenith during the interwar period. Modernist poets of the time included E. E. Cummings and Wallace Stevens, while non-modernist authors included Theodore Dreiser, Ernest Hemingway, Scott Fitzgerald, and John Steinbeck. Hugh MacDiarmid, a Scottish poet, was a notable British author in the 1920s and 1930s, alongside Virginia Woolf, E. M. Forster, Evelyn Waugh, and P.G. Wodehouse.

Aldous Huxley's dystopian masterwork Brave New World was published in 1932. Tropic of Cancer, Henry Miller's controversial book on his experiences as a struggling author in Paris, was published two years later, in 1934. Graham Greene's Brighton Rock and James Joyce's Finnegans Wake were also published in 1938.

- *Post World War II Literature*

Soon after World War II, a revived craving for spiritual connection and religious conviction emerged as a basic and uniting literary theme, as indicated by works by such divergent authors and poets as W.H. Auden, T.S. Eliot, Evelyn Waugh, and Christopher Fry.

- *Childrens Literature*

As the 20th century dawned, children's literature was primarily designed to teach moral lessons and instill traditional values. However, the rise of consumer culture and mass media, including radio, television, and film, challenged the traditional roles of books in children's lives. Publishers began to experiment with new forms of storytelling and illustration, while writers began to tackle more complex and controversial themes.

The early 20th century saw the emergence of classic works of children's literature, such as L. Frank Baum's "The Wonderful Wizard of Oz" and E.B. White's "Charlotte's Web," which addressed issues of identity, friendship, and social justice. The Great Depression and World War II brought further changes to children's literature, as writers sought to provide comfort and hope to young readers in difficult times.

The post-war years were marked by the emergence of new literary forms and styles, including the picture book and the young adult novel. Children's literature also became more diverse and inclusive, with writers and illustrators from different backgrounds and cultures sharing their stories with young readers.

- *The Novel: the 1940s onwards*

The period spanning from the 1940s onwards stands as a testament to the transformative power of the novel as a literary form. During this time, novelists reached new heights of creativity and insight, tackling complex societal issues, exploring the human psyche, and pushing the boundaries of narrative structure and style.

From the post-war era to the close of the century, novelists grappled with the changing social, political, and cultural landscape, reflecting the profound impact of global events such as World War II and the Cold War, the civil rights movement, the rise of feminism, and the onset of postmodernism. Through their works, they explored dystopian futures, the collapse of social order, the inner workings of the mind, the complexities of identity, and the essence of human nature. Whether through the vivid social panorama of Powell's "A Dance to the Music of Time," the totalitarian nightmares of Orwell's "1984" and Burgess's "A Clockwork Orange," the psychological insights of Golding's "Lord of the Flies" and McEwan's "The Cement Garden," or the experimental narrative techniques of Gray's "Lanark," these novels not only reflect the era's literary innovation but also continue to resonate with contemporary readers.

In Britain, authors like Anthony Powell, George Orwell, William Golding, Muriel Spark, Anthony Burgess, Richard Adams, Ian McEwan, and Alasdair Gray produced works that have since become classics, each contributing uniquely to the literary tapestry of the time. They captured the zeitgeist of the era, offering readers immersive, thought-provoking explorations of life and society in the 20th century.

The depth and diversity of these works underscore the novel's enduring power as a vehicle for social critique, philosophical inquiry, and humanistic expression, marking the period from the 1940s onwards as a pivotal chapter in the history of 20th-century literature.

Printed in Great Britain
by Amazon

dfe76e56-bad2-4f03-af3f-939e370711e3R01